Bayesian Analysis
of Time Series

Bayesian Analysis of Time Series

Lyle D. Broemeling

CRC Press
Taylor & Francis Group
Boca Raton London New York

CRC Press is an imprint of the
Taylor & Francis Group, an **Informa** business

CRC Press
Taylor & Francis Group
6000 Broken Sound Parkway NW, Suite 300
Boca Raton, FL 33487-2742

First issued in paperback 2021

Version Date: 2019

ISBN-13: 978-0-367-77999-3 (pbk)
ISBN-13: 978-1-138-59152-3 (hbk)

Library of Congress Cataloging-in-Publication Data

Names: Broemeling, Lyle D., 1939- author.
Title: Bayesian analysis of time series / Lyle D. Broemeling.
Description: Boca Raton : CRC Press, Taylor & Francis Group, 2019. | Includes bibliographical references
Identifiers: LCCN 2018061555 | ISBN 9781138591523
Subjects: LCSH: Time-series analysis–Textbooks. | Bayesian statistical decision theory—Textbooks
Classification: LCC QA280 .B765 2019 | DDC 519.5/5–dc23
LC record available at https://lccn.loc.gov/2018061555

Visit the Taylor & Francis Web site at
http://www.taylorandfrancis.com

and the CRC Press Web site at
http://www.crcpress.com

Contents

About the Author

Lyle D. Broemeling, Ph.D., is Director of Broemeling and Associates Inc., and is a consulting biostatistician. He has been involved with academic health science centers for about 20 years and has taught and been a consultant at the University of Texas Medical Branch in Galveston, The University of Texas MD Anderson Cancer Center and the University of Texas School of Public Health. His main interest is in developing Bayesian methods for use in medical and biological problems and in authoring textbooks in statistics. His previous books are Bayesian Biostatistics and Diagnostic Medicine, and Bayesian Methods for Agreement.

1

Introduction to the Bayesian Analysis of Time Series

1.1 Introduction

This book presents the Bayesian approach to the analysis of time series and illustrates the techniques related to this with many examples. Using the software packages R and WinBUGS enhances the way a Bayesian approach can be implemented. The R package is primarily used to generate observations from a given time series model, while the WinBUGS package allows one to perform a posterior analysis that provides the way to determine the characteristic of the posterior distribution of the unknown parameters. It is important to realize that the reader most likely would not execute the Bayesian analysis provided by the author. This is because the investigator would have different prior information and possibly a different likelihood for a given sample information. Many of the examples are based on data generated by R and the author places noninformative prior distribution on the parameters. It is also important to note that the posterior analysis executed by WinBUGS would also most likely be different than the analyses that appear in the book. This is due to the fact that others would possibly use a different number of observations for the Monte Carlo Markov Chain (MCMC) simulation with different prior distributions and different initial values for the parameters. There are not many books on time series from a Bayesian point of view, but the author has based his approach to some extent on earlier textbooks such as Pole and West,[1] West and Harrison,[2] Spall,[3] and Venkatesan and Aramugam.[4] I have also relied heavily on the books by Broemeling[5,6] that present some of the earliest ideas on Bayesian time series.

My approach in this book is presented in roughly three phases: (1) First is the data that are sometimes generated by R from a time series whose parameters are known or data from a scientific investigation, (2) a specification of the model and prior distribution of the unknown parameters, and (3) the posterior analysis either presented analytically or numerically by WinBUGS. My book is somewhat unique in that it uses R and WinBUGS to expedite the analysis of a given series.

The rest of the chapter will preview the chapters of the book.

1.2 Chapter 2: Bayesian Inference and MCMC Simulation

The chapter begins with a description of Bayes theorem followed by the four phases of statistical inference, namely: (1) prior information, (2) sample information represented by the likelihood function, (3) the joint posterior distribution of the parameters, and (4) if appropriate, forecasting future observations via the Bayesian predictive density (or mass function). Bayesian inferential techniques are illustrated with many examples. For example, the first population to be considered is the binomial with one parameter, the probability of success in n independent trials, where a beta prior is placed on the parameter, which when combined with the likelihood function gives a beta posterior distribution for the probability of success. Based on the posterior distribution of the parameter, point estimation, and interval estimation are described, and finally the Bayesian predictive mass function is derived for m future independent trials, and demonstrations for the binomial population is done with $n=11$ trials and $m=5$ future trials. The same four-phase scenario is explained for the normal, multinomial, and Poisson populations.

The remaining sections of the chapter are devoted to the description of MCMC algorithms for the simulation of observations from the relevant posterior distributions. Also included in the concluding sections of the chapter are detailed directions for how to use R and WinBUGS.

1.3 Chapter 3: Fundamentals of Time Series Analysis

A time series is defined as a stochastic process whose index is time, and several examples are explained:

(1) Airline passenger bookings over a 10-year period, (2) the sunspot cycle, and (3). Annual Los Angeles rainfall data, starting in 1903. The first two series have an increasing trend over time and clearly have seasonal components, while the third (the rainfall data) has a constant trend and no obvious seasonal effects. At this point, the student is shown how to employ R to graph the data and how to delineate the series into trend and seasonal components. The R package has a function called decomposition that portrays the observations, the trend of the series, the seasonal component, and the errors of a given series. Decomposition is a powerful tool and should always be used when analyzing a time series. This will give the investigator enough information to propose a tentative time series model. Also presented are definitions of the mean value function and variance function of the time series, the autocorrelation function, and the correlogram. These characteristic of a time series are computed for the airline passenger booking data, the sunspot cycle data, and the Los

Angeles rainfall data using the appropriate R command (e.g. mean(), std(), autocorrelation or acf(), and partial autocorrelation or pacf()).

1.4 Chapter 4: Basic Random Models

This chapter describes the building blocks for constructing more complex time series models, beginning with white noise, a sequence of independent normally identically distributed random variables. This sequence will serve as the error terms for the autoregressive time series models. We now progress from white noise to the random walk time series, and random walk with drift. For the autoregressive time series, the likelihood function is explicated, a prior distribution for the parameters is specified, and via Bayes theorem, the posterior distribution of the autoregressive parameters and precision of the error terms are derived. R is employed to generate observations from an AR(1) model with known parameters; then using those observations as data, WinBUGS is executed for the posterior analysis which gives Bayesian point and interval estimates of the unknown parameters. Since the parameter values used to generate the data are known, the posterior means and medians can be compared to them giving one some idea of how 'close' these estimates are to the so-called 'true' values. Another example of an autoregressive time series is presented, followed by a Bayesian approach to goodness-of-fit, and finally the Bayesian predictive density for a future observation is derived.

1.5 Time Series and Regression

The first time series to be presented is the linear regression model with AR (1) errors, where R is employed to generate observations from that linear model with known parameters followed by an execution with WinBUGS that produces point and interval estimates of those parameters. The scenario is repeated for a quadratic regression with seasonal effects and AR(1) errors. In all scenarios, R code generates the data and using that as information for the Bayesian analysis, the posterior distribution of the parameters is easily provided. Various generalizations to more complicated situations include a nonlinear regression model (with exponential trend) with AR(1) errors, a linear regression model with AR(2) errors which contains five parameters, two linear regression coefficients, two autoregressive parameters, plus the precision of the error terms. An interesting time series model is one which is continuous time, which is the solution to a stochastic differential equation, where the posterior distribution of the autoregressive parameter is derived assuming a noninformative prior distribution for the autoregressive parameter and the variance of the errors. The chapter concludes with a section on comments and conclusions followed by eight problems and eight references.

1.6 Time Series and Stationarity

Stationarity of a time series is defined including stationary in the mean and stationary in the variance. The first time series to be considered is the moving average model and the first and second moments are derived. It is important to note that moving average processes are always stationary. R is employed to generate observations from an MA(1) series with known parameters (the moving average coefficient and the precision of the white noise); then using those observations, WinBUGS is executed for the posterior analysis which provides point and credible intervals for the two parameters. The MA(1) series is generalized to quadratic regression model with MA(1) errors, where again R generates observations from it where the parameters are known.

Bayesian analysis via WinBUGS performs the posterior analysis. An additional generalization focuses on a quadratic regression model with an MA(2) series for the errors with R used to generate observations from the model but with known values for the parameters. It is interesting to compare the results of the posterior estimates of the quadratic regression with MA(1) errors to that with MA(1) errors. Generalizing to a regression model with a quadratic trend, but also including seasonal effects, presents a challenge to the Bayesian way of determining the posterior analysis. The predictive density for forecasting future observations is derived, and the last section of the chapter presents the Bayesian technique of testing hypotheses about the moving average parameters. There are 16 exercises and 5 references.

1.7 Time Series and Spectral Analysis

It is explained how spectral analysis gives an alternative approach to studying time series, where the emphasis is on the frequency of the series, instead of the time. Chapter 7 continues with a brief history of the role of the spectral density function in time series analysis, and the review includes an introduction to time series with trigonometric components (sine and cosine functions) as independent variables. Next to be explained are time series with trigonometric components whose coefficients are Fourier frequencies and their posterior distributions are revealed with a Bayesian analysis. It is important to remember that frequency is measured in hertz (one cycle per second) or in periods measured as so many units of pi radians per unit time. It can be shown that for the basic time series (autoregressive and moving average series), the spectral density function is known in terms of the parameters of the corresponding model. It is important to remember the spectrum is related to the Fourier line spectrum which is a plot of the fundamental frequencies versus m, where m is the number of harmonics in the series. Section 10 continues with the derivation of the spectral density function for the basic series, AR(1), AR(2), MA(1), MA(2), ARMA(1,1), etc. Remember that the spectral density function is a function of frequency (measured in hertz or multiples of pi in radians per cycle). For each model, R is used to generate observations from that model (with known parameters); then, using those observations as data, the Bayesian analysis is executed with WinBUGS and as a consequence the posterior distribution of the spectral density at various frequencies is estimated. For the last example, the spectral density of the sunspot cycle is estimated via a Bayesian analysis executed with WinBUGS. There are 17 problems followed by y references.

1.8 Dynamic Linear Model

The dynamic linear model consists of two equations: the observation equation and the system equation.

The first expresses the observations as a linear function of the states of the system, while the second expresses the current state as a linear function of previous states. Usually, the system equation is a first-order autoregressive series, and various expressions of the dynamic linear model depend on what is assumed about the parameters of the model. First to be considered is the dynamic model in discrete time and the various phases of inference are explained. The various phases are as follows: (1) estimation, (2) filtering, (3) smoothing, and (4) prediction. Filtering is the posterior distribution of the present state at time t induced by the prior distribution of the previous state at time $t - 1$ in a sequential manner, and this recursive Bayesian algorithm is called the Kalman Filter. An important generalization of the dynamic linear model is where the system equation is amended to include a control variable, which is quite useful in navigation problems. Briefly, the control problem is as follows: at time t, the position of the object (satellite, missile, etc.) is measured as the observation at t, then the next position of the object is predicted so that the posterior mean of the next state is equal to the target value (the desired position) at time $t + 1$. These ideas are illustrated with various examples beginning with a model where the observation and system vectors are univariate with known coefficients. For this case, the Kalman filter estimates the current state in a recursive manner as described earlier. As usual, R generates the observations and states of the system via the observation and system equations of the dynamic linear model, which is followed by a Bayesian analysis via WinBUGS code that estimates the current state (the Kalman filter) of the system. R includes a powerful package dlm that will generate the observations and states of the process. One must download the TSA package and dlm package into R! The last section of the chapter deals with the control problem, where the choice of the control variable is described by a six-step algorithm. Finally, the problem of adaptive estimation is explained; this a generalization of the dynamic linear model, where the precision parameters of the observation and system equations are unknown. An example involving bivariate observations and states and unknown 2×2 precision matrices illustrates the Bayesian approach to the Kalman filter. The bivariate observations and states are generated with the dlm R package. There are 14 exercises and 21 references.

1.9 The Shift Point Problem

Shift point models are defined in general, followed by an example involving a univariate normal sequence of normal random variables with one shift where the time when the shift occurs is known. For such a case, the posterior distribution of the parameters of the two sequences is derived, assuming noninformative prior distributions for the parameters. There are three parameters: the mean of the first and second phases, plus the common precision parameter of both. Next to be considered is a two-phase AR(1) model with one shift but unknown time where the shift occurs. R generates the observations from this model where the time when the shift occurs is known but where the shift point is a random variable. A prior distribution is specified for the distribution of the shift point (where the shift occurs), the common precision of the errors, and the two AR(1) parameters. Of course, the parameters of the model are assumed to be known when generating the observations via R, and the posterior analysis executed with WinBUGS. The next series to be considered with one shift is the MA(1) and a similar scenario of generating observations via R and posterior analysis via WinBUGS. A large part of the chapter focuses on examples taken from econometrics. For example, a regression model with AR(1) errors and one gradual shift (modeled with a transition function) is presented and the posterior distribution of the continuous shiftpoint is derived. This is repeated for a linear regression model with MA(1) errors and one unknown shift point, followed by a section on testing hypotheses concerning the moving average parameter of the two phases. An important subject in shift point models is that of threshold autoregressive models, where the model changes parameters depending on when the observation exceeds a given value, the threshold. The chapter ends with 10 problems and 28 references.

1.10 Chapter 10: Residuals and Diagnostic Tests

Residuals of a time series model are defined and their use as diagnostic checks for how well the model fits the data is described. Procedures for calculating residuals and the corresponding standardized residuals are explained followed by many examples that illustrate those procedures. R generates observations from an AR(1) model with known parameters, and using these as data, a Bayesian analysis computes the posterior characteristics of the residuals. Given the posterior means of the residuals, the sample mean and standard deviation of the residuals are computed. Based on the mean and standard deviation of the residuals, the posterior analysis is repeated but this time the posterior means of the standardized residuals are computed. A normal q–q plot of the standardized residuals is evaluated to see how well the model fits the data. If the model fits the data well, the normal q–q plot should appear as linear. The more pronounced the linearity, the more confident is one of the models fit. The above scenario of diagnostic checks is repeated for a linear regression model with AR(1) errors, and a linear regression model with MA(1) errors.

The chapter concludes with comments and conclusions, four problems, and eight references.

It is hoped that this brief summary of the remaining chapters will initiate enough interest in the reader so that they will continue to learn how the Bayesian approach is used to analyze time series.

References

1. Pole, A. and West, M. (1994). *Applied Bayesian Forecasting and Time Series Analysis*. Taylor and Francis, Boca Raton, FL.
2. West, M. and Harrison, J. (1997). *Bayesian Forecasting and Dynamic Models, Second Edition*. Springer, New York, NY.
3. Spall, J. (1988). *Bayesian Analysis of Time Series and Dynamic Models*. Marcel-Dekker Inc., New York, NY.
4. Venkatesan, D. and Aramugam, P. (2011). *A Bayesian Analysis of Changing Time Series Models: A Monograph on Time Series Analysis*. CRC Press, Taylor & Francis, Boca Raton, FL.
5. Broemeling, L.D. (1985). *Bayesian Analysis of Linear Models*. Marcel-Dekker Inc., New York, NY.
6. Broemeling, L.D. (1987). *Econometric and Structural Change*. Marcel-Dekker Inc., New York, NY.

2

Bayesian Analysis

2.1 Introduction

Bayesian methods will be employed to make Bayesian inferences for time series, and this chapter will introduce the theory that is necessary in order to describe those Bayesian procedures. Bayes theorem, the foundation of the subject, is first introduced and then followed by an explanation of the various components of Bayes theorem: prior information, information from the sample given by the likelihood function, the posterior distribution which is the basis of all inferential techniques, and finally the Bayesian predictive distribution. A description of the main three elements of inference, namely, estimation, tests of hypotheses, and forecasting future observations follows.

The remaining sections refer to the important standard distributions for Bayesian inference, namely, the Bernoulli, the beta, the multinomial, Dirichlet, normal, gamma, and normal-gamma, multivariate normal, Wishart, normal-Wishart, and the multivariate t-distributions. As will be seen, the relevance of these standard distributions to inferential techniques is essential for understanding the analysis of time series.

As will be seen, the multinomial and Dirichlet are the foundation for the Bayesian analysis of Markov chains and Markov jump processes. For normal stochastic processes such as Wiener and Brownian motion, the multivariate normal and the normal-Wishart play a key role in determining Bayesian inferences.

Of course, inferential procedures can only be applied if there is adequate computing available. If the posterior distribution is known, often analytical methods are quite sufficient to implement Bayesian inferences, and will be demonstrated for the binomial, multinomial, and Poisson populations, and several cases of normal populations. For example, when using a beta prior distribution for the parameter of a binomial population, the resulting beta posterior density has well-known characteristics, including its moments. In a similar fashion, when sampling from a normal population with unknown mean and precision and with a vague improper prior, the resulting posterior *t*-distribution for the mean has known moments and percentiles which can be used for inferences.

Posterior inferences by direct sampling methods are easily done if the relevant random number generators are available. However, if the posterior distribution is quite complicated and not recognized as a standard distribution, other techniques are needed. To solve this problem, Monte–Carlo–Markov Chain (MCMC) techniques have been developed and have been a major success in providing Bayesian inferences for quite complicated problems. This has been a great achievement in the field and will be described in later sections.

Minitab, S-Plus, WinBUGS, and R are packages that provide random number generators for direct sampling from the posterior distribution for many standard distributions, such as binomial, gamma, beta, and *t*-distributions. On occasion, these will be used; however, my preferences are WinBUGS and R, because they have been adopted by other Bayesians. This is also true for indirect sampling, where WinBugs and R are excellent packages and are preferred for this book. Many institutions provide special purpose software for specific Bayesian routines. For example, at MD Anderson Cancer Center, where Bayesian applications are routine, several special purpose programs are available for designing (including sample size justification) and analyzing clinical trials, and will be described. The theoretical foundation for MCMC is introduced in the following sections.

Inferences for time series consist of testing hypotheses about unknown population parameters, estimation of those parameters, and forecasting future observations. When a sharp null hypothesis is involved, special care is taken in specifying the prior distribution for the parameters. A formula for the posterior probability of the null hypothesis is derived, via Bayes theorem, and illustrated for Bernoulli, Poisson, and normal populations. If the main focus is estimation of parameters, the posterior distribution is determined, and the mean, median, standard deviation, and credible intervals found, either analytically or by computation with WinBUGS or R. For example, when sampling from a normal population with unknown parameters and using a conjugate prior density, the posterior distribution of the mean is a *t* and will be derived algebraically. However, for making Bayesian inferences for time series such as autoregressive are univariate *t* for the individual coefficients are gamma for the precision of the error process of the time series. These posterior inferences are provided both analytically and numerically with WinBUGS or R. Of course, all analyses should be preceded by checking to determine if the model is appropriate, and this is where the predictive distribution comes into play. By comparing the observed results of the experiment with those predicted, the model assumptions are tested. The most frequent use of the Bayesian predictive distribution is for forecasting future observation of time series.

2.2 Bayes Theorem

Bayes theorem is based on the conditional probability law:

$$P[A|B] = P[B|A]P[A]/P[B] \tag{2.1}$$

where $P[A]$ is the probability of A before one knows the outcome of the event B, $P[B|A]$ is the probability of B assuming what one knows about the event A, and $P[A|B]$ is the probability of A knowing that event B has occurred. $P[A]$ is called the prior probability of A, while $P[A|B]$ is called the posterior probability of A.

Another version of Bayes theorem is to suppose X is a continuous observable random vector and $\theta \in \Omega \subset R^m$ is an unknown parameter vector, and suppose the conditional density of X given θ is denoted by $f(X|\theta)$. If $X = (X_1, X_2, \ldots, X_n)$ represents a random sample of size n from a population with density $f(X|\theta)$, and $\xi(\theta)$ is the prior density of θ, then Bayes theorem expresses the posterior density as

$$\xi(\theta|X) = C \prod_{i=1}^{i=} f(x_i|\theta)\xi(\theta), \quad X_i \in R \text{ and } \theta \in \Omega, \tag{2.2}$$

where the proportionality constant is c, and the term $\prod_{i=1}^{i=n} f(x_i|\theta)$ is called the likelihood function. The density $\xi(\theta)$ is the prior density of θ and represents the knowledge one possesses about the parameter before one observes X. Such prior information is most likely available to the experimenter from other previous related experiments. Note that θ is considered a random variable and that Bayes theorem transforms one's prior knowledge of θ, represented by its prior density, to the posterior density, and that the transformation is the combining of the prior information about θ with the sample information represented by the likelihood function.

'An essay toward solving a problem in the doctrine of chances' by the Reverend Thomas Bayes[1] is the beginning of our subject. He considered a binomial experiment with n trials and assumed the probability θ of success was uniformly distributed (by constructing a billiard table) and presented a way to calculate $\Pr(a \leq \theta \leq b | X = p)$, where X is the number of successes in n independent trials. This was a first in the sense that Bayes was making inferences via $\xi(\theta|X)$, the conditional density of θ given x. Also, by assuming the parameter as uniformly distributed, he was assuming vague prior information for θ. The type of prior information where very little is known about the parameter is called noninformative or vague information.

It can well be argued that Laplace[2] is the greatest Bayesian because he made many significant contributions to inverse probability (he did not know of Bayes), beginning in 1774 with 'Memorie sur la probabilite des causes par la evenemens,' with his own version of Bayes theorem, and over a period of some 40 years culminating in 'Theorie analytique des probabilites.' See Stigler[3] and Chapters 9–20 of Hald[4] for the history of Laplace's contributions to inverse probability.

It was in modern times that Bayesian statistics began its resurgence with Lhoste,[5] Jeffreys,[6] Savage,[7] and Lindley.[8] According to Broemeling and Broemeling,[9] Lhoste was the first to justify noninformative priors by invariance principals, a tradition carried on by Jeffreys. Savage's book was a major contribution in that Bayesian inference and decision theory was put on a sound theoretical footing as a consequence of certain axioms of probability and utility, while Lindley's two volumes showed the relevance of Bayesian inference to everyday statistical problems and was quite influential and set the tone and style for later books such as Box and Tiao,[10] Zellner,[11] and Broemeling.[12] Books by Box and Tiao and Broemeling were essentially works that presented Bayesian methods for the usual statistical problems of the analysis of variance and regression, while Zellner focused Bayesian methods primarily on certain regression problems in econometrics. During this period, inferential problems were solved analytically or by numerical integration. Models with many parameters (such as hierarchical models with many levels) were difficult to use because at that time numerical integration methods had limited capability in higher dimensions. For a good history of inverse probability, see Chapter 3 of Stigler,[3] and Hald,[4] who present a comprehensive history and are invaluable as a reference..

The last 20 years are characterized by the rediscovery and development of resampling techniques, where samples are generated from the posterior distribution via MCMC methods, such as Gibbs sampling. Large samples generated from the posterior make it possible to make statistical inferences and to employ multi-level hierarchical models to solve complex, but practical problems. See Leonard and Hsu,[13] Gelman et al.,[14] Congdon,[15–17] Carlin and Louis,[18] Gilks, Richardson, and Spiegelhalter,[19] who demonstrate the utility of MCMC techniques in Bayesian statistics.

2.3 Prior Information

2.3.1 The Binomial Distribution

Where do we begin with prior information, a crucial component of Bayes theorem rules? Bayes assumed the prior distribution of the parameter is uniform, namely

$\xi(\theta) = 1, 0 \le \theta \le 1$, where

θ is the common probability of success in n independent trials and

$$f(x|\theta) = \binom{n}{x}\theta^x(1-\theta)^{n-x}, \tag{2.3}$$

where x is the number of successes $x = 0, 1, 2, \ldots, n$. For the distribution of X, the number of successes is binomial and denoted by $X \sim \text{Binomial}(\theta, n)$. The uniform prior was used for many years; however, Lhoste[5] proposed a different prior, namely

$$\xi(\theta) = \theta^{-1}(1-\theta)^{-1}, \ 0 \le \theta \le 1, \tag{2.4}$$

to represent information which is noninformative and is an improper density function. Lhoste based the prior on certain invariance principals, quite similar to Jeffreys.[6] Lhoste also derived a noninformative prior for the standard deviation σ of a normal population with density

$$f(x|\mu, \sigma) = (1/\sqrt{2\pi}\sigma)\exp-(1/2\sigma)(x-\mu)^2, \mu \in R \text{ and } \sigma > 0. \tag{2.5}$$

He used invariance as follows: he reasoned that the prior density of σ and the prior density of $1/\sigma$ should be the same, which leads to

$$\xi(\sigma) = 1/\sigma. \tag{2.6}$$

Jeffreys' approach is similar in that in developing noninformative priors for binomial and normal populations, but he also developed noninformative priors for multi-parameter models, including the mean and standard deviation for the normal density as

$$\xi(\mu, \sigma) = 1/\sigma, \ \mu \in R \text{ and } \sigma > 0. \tag{2.7}$$

Noninformative priors were ubiquitous from the 1920s to the 1980s and were included in all the textbooks of that period. For example, see Box and Tiao,[10] Zellner,[11] and Broemeling.[12] Looking back, it is somewhat ironic that noninformative priors were almost always used, even though informative prior information was almost always available. This limited the utility of the Bayesian approach, and people saw very little advantage over the conventional way of doing business. The major strength of the Bayesian way is that it a convenient, practical, and logical method of utilizing informative prior information. Surely the investigator knows informative prior information from previous related studies.

How does one express informative information with a prior density? For example, suppose one has informative prior information for the binomial population. Consider

$$\xi(\theta) = [\Gamma(\alpha+\beta)/\Gamma(\alpha)\Gamma(\beta)]\theta^{\alpha-1}(1-\theta)^{\beta-1}, \quad 0 \le \theta \le 1, \qquad (2.8)$$

as the prior density for θ. The beta density with parameters α and β has mean $[\alpha/(\alpha+\beta)]$ and variance $[\alpha\beta/(\alpha+\beta)^2(\alpha+\beta+1)]$ and can express informative prior information in many ways.

As for prior information for the binomial, consider the analysis of Markov processes, namely in estimating the transition probability matrix of a stationary finite-state Markov chain. Consider the five-by-five transition matrix P with components
$P = (p_{ij})$, where

$$P = \begin{pmatrix} .2,,2,.2,.2,.2 \\ .2,.2,.2,.2,.2 \\ .2,.2,.2,.2,.2 \\ .2,.2,.2,.2,.2 \\ .2,.2,.2,.2,.2 \end{pmatrix}. \qquad (2.9)$$

Note that

$$p_{ij} = \Pr[X_{n+1} = j | X_n = i] \qquad (2.10)$$

where $n = 0, 1, 2, \ldots$

that is to say X_n is a discrete time Markov with state-space

$S = \{1, 2, 3, 4, 5\}$. Note that p_{ij} are the one-step transition probabilities of the Markov chain X_n, where the first row is the conditional distribution (given $X_0 = 1$) of a discrete random variable with mass points 1, 2, 3, 4, and 5, with probabilities 0.2, 0.2, 0.2, 0.2, 0.2. The second row is the conditional distribution (given $X_0 = 2$) of a discrete random variable with mass points 1, 2, 3, 4, and 5 with probabilities 0.2, 0.2, 0.2, 0.2, 0.2, etc. This is an example of a Markov chain where each state is recurrent, that is, it is possible to reach any state from any other state,

Using a multinomial distribution with probability mass function

$$\prod_{i,j=1}^{i,j=5} p_{ij}^{n_{ij}}, \qquad (2.11)$$

where $\sum_{i,j=1}^{i,j=5} p_{ij} = 1$ and $\sum_{i,j=1}^{i,j=5} n_{ij} = n$.

98 n_{ij} values are generated from the chain with the result

1 2 5 1 4 1 5 3 3 2 4 4 2 5 5 5 3 2 2 2 3 2 1 4 5 3 2 1 2 3 4 3 3 4 2 4 5 1 5 1 1 4
5 1 4 1 3 4 4 2 5 1 5 1 3 1 2 4 1 3 1 3 5 5 4 1 4 1 2 1 2 5 4 4 4 2 4 2 5 5 4 5 2 4 3
2 2 3 2 1 2 1 2 4 2 3 1 4 4 2.

The R code below is used to simulate the 98 observations from the Markov chain with transition matrix P

RC 2.1

```
MC.sim<-function(n,P,x1){
sim<-as.numeric(n)
m<-ncol(P)
if (missing(x1)){
sim[1]<-sample(1:m,1)# random start
} else {sim[1]<-x1}
for ( i in 2:n){
newstate<-sample(1:m,1,prob=P[sim[i-1],])
sim[i]<-newstate
}
sim
}
P<-matrix(c(.2,.2,.2,.2,.2,.2,.2,.2,.2,.2,.2,.2,.2,.2,.2,.2,.2,.2,.2,.2,.2,.2,.2,.2,.2),
nrow=5,ncol=5,byrow=TRUE)
MC.sim(100,P,1)
```

These cell frequencies can be displayed with the 5 by 5 matrix

$$N = \begin{pmatrix} 1,7,4,6,2 \\ 5,3,4,6,5 \\ 3,6,2,3,1 \\ 5,7,2,5,4 \\ 6,1,3,3,4 \end{pmatrix} \qquad (2.12)$$

Thus, there are one one-step transition from 1 to 1, seven transitions from 1 to 2, and finally two one-step transitions from 1 to 5. Since the simulation was based on the multinomial distribution, it is know that the marginal distribution of the cell frequency n_{ij} is binomial with parameters p_{ij} and $n=98$. In order to perform a Bayesian analysis, a prior distribution is assigned to the unknown cell frequencies: The conjugate distribution to the multinomial is the Dirichlet, which induces a beta prior to the individual cell frequencies. This results in a Dirichlet for the posterior distribution of the transition probabilities p_{ij}, and consequently a beta for the individual transition probabilities. For the Dirichlet, the density is

$$f(p_{11}, p_{12}, \ldots, p_{55}) \propto \prod_{i,j=1}^{i,j=5} p_{ij}^{\alpha_{ij}-1}, \qquad (2.13)$$

where $\sum_{i,j=1}^{i,j=5} p_{ij} = 1$ and the α_{ij} are positive.

Later, in this chapter, a posterior analysis for estimating the transition probabilities will be presented.

2.3.2 The Normal Distribution

Of course, the normal density plays an important role as a model for time series. For example, as will be seen in future chapters, the normal distribution will model the observations of certain time series, such as autoregressive and moving average series. How is informative prior information expressed for the parameters μ and σ (the mean and standard deviation)? Suppose a previous study has m observations $x = (x_1 x_2, \ldots, x_m)$, then the density of X given μ and σ is

$$f(x \mid \mu, \sigma) \propto \left[\sqrt{m} / \sqrt{2\pi\sigma^2} \right] \exp - (m/2\sigma^2)(\bar{x} - \mu)^2$$

$$\left[(2\pi)^{-(n-1)/2} \sigma^{-(n-1)} \right] \exp - (1/2\sigma^2) \sum_{i-1}^{i=m} (x_i - \bar{x})^2 \tag{2.14}$$

This is a conjugate density for the two-parameter normal family and is called the normal-gamma density. Note it is the product of two functions, where the first, as a function of μ and σ, is the conditional density of μ given σ, with mean $\bar{x}$ and variance σ^2/m, while the second is a function of σ only and is an inverse gamma density. Or equivalently, if the normal is parameterized with μ and the precision $\tau = 1/\sigma^2$, the conjugate distribution is as follows: (a) the conditional distribution of μ given τ is normal with mean $\bar{x}$ and precision $m\tau$, and (b) the marginal distribution of τ is gamma with parameters $(m+1)/2$ and $\sum_{i=1}^{i=m} (x_i - \bar{x})^2/2 = (m - 1)S^2/2$, where S^2 is the sample variance. Thus, if one knows the results of a previous experiment, the likelihood function for μ and τ provides informative prior information for the normal population.

For example, the normal serves as the distribution of the observations of a first-order autoregressive process

$$Y(t) = \theta Y(t - 1) + W(t), t = 1, 2, \ldots. \tag{2.15}$$

where

$$W(t), t = 1, 2, \ldots \tag{2.16}$$

is a sequence of independent normal random variables with mean zero and precision τ, and $\tau > 0$. It is easy to show that the joint distribution of the n observations from the AR(1) process is multivariate normal with mean vector

0 and variance covariance matrix with diagonal entries $1/\tau(1 - \theta^2)$ and k-th order covariance $Cov[Y(t), Y(t+k)] = \theta^k/\tau(1 - \theta^2), |\theta| < 1, k = 1, 2, ...$

Note it is assumed the process is stationary, namely, $|\theta| < 1$. Of course, the goal of the Bayesian analysis is to estimate the processes autoregressive parameter θ and the precision $\tau > 0$. For the Bayesian analysis, a prior distribution must be assigned to θ and τ, which in the conjugate prior case is a normal-gamma. The posterior analysis for the autoregressive time series results in a univariate t-distribution for the distribution of θ as will be shown in Chapter 5.

2.4 Posterior Information

2.4.1 The Binomial Distribution

The preceding section explains how prior information is expressed in an informative or in a noninformative way. Several examples are given and will be revisited as illustrations for the determination of the posterior distribution of the parameters. Suppose a uniform prior distribution for the transition probability (of the five-state Markov chain) p_{ij} is used. What is the posterior distribution of p_{ij}?

By Bayes theorem,

$$f(p_{ij}|N) \propto \binom{n}{n_{ij}} p_{ij}^{n_{ij}} (1 - p_{ij})^{n-n_{ij}} \tag{2.17}$$

where n_{ij} is the observed transitions from state i to state j and n is the total cell counts for the five-by-five cell frequency matrix N. Of course, this is recognized as a beta $(n_{ij} + 1, n - n_{ij} + 1)$ distribution, and the posterior mean is $(n_{ij} + 1/n + 2)$. However, if the Lhoste[5] prior density (2.4) is used, the posterior distribution of p_{ij} is beta $(n_{ij}, n - n_{ij})$ with mean n_{ij}/n, which is the usual estimator of p_{ij}.

2.4.2 The Normal Distribution

Consider a random sample $X = (x_1, x_2, ..., x_n)$ of size n from a normal $(\mu, 1/\tau)$ population, where $\tau = 1/\sigma^2$ is the inverse of the variance, and suppose the prior information is vague and the Jeffreys–Lhoste prior $\xi(\mu, \tau) \propto 1/\tau$ is appropriate, then the posterior density of the parameters is

$$\xi(\mu, \tau|\text{data}) \propto \tau^{n/2-1} \exp -(\tau/2)[n(\mu - \bar{x})^2 + \sum_{i=1}^{i=n} (x_i - \bar{x})^2]. \tag{2.18}$$

Using the properties of the gamma density, τ is eliminated by integrating the joint density with respect to τ to give

$$\xi(\mu|\text{data}) \propto \qquad (2.19)$$

$$\left\{\Gamma(n/2)n^{1/2}/(n-1)^{1/2}S\pi^{1/2}\Gamma((n-10/2)\right\}/[1+n(\mu-\bar{x})^2/(n-1)S^2]^{(n-1+1)/2}$$

which is recognized as a t-distribution with $n-1$ degrees of freedom, location $\bar{x}$, and precision n/S^2. Transforming to $(\mu-\bar{x})\sqrt{n}/S$, the resulting variable has a Student's t-distribution with $n-1$ degrees of freedom. Note the mean of μ is the sample mean, while the variance is $[(n-1)/n(n-3)], n>3$.

Eliminating μ from (12) results in the marginal distribution of τ as

$$\xi(\tau|s^2) \propto \tau^{[(n-1)/2]-1} \exp{-\tau(n-1)S^2/2}, \ \tau>0, \qquad (2.20)$$

which is a gamma density with parameters $(n-1)/2$ and $(n-1)s^2/2$. This implies the posterior mean is $1/s^2$ and the posterior variance is $2/(n-1)s^4$.

For example, consider the AR(1) (2.15) series where $\theta=.6$ and $\sigma^2=1$, then suppose R is used to generate a realization of $n=50$ from the series.

Our goal is to determine the posterior distribution of θ and $\tau=1/\sigma^2$.

The R code used for the simulation is given below, namely:

RC 2.2

```
set.seed(1)
y<-w<-rnorm(50)
for ( t in 2:50){y[t]<-.6*y[t-1]+w[t]
time<-1:50
plot(time,y)
```

The vector y contains the 50 simulated values for the autoregressive time series.

Y=(−.62,−.1922,−.9509,1.0247,.9443,−.2538,.3351,.93938,1.13954,.37826, 1.73873,1.43308,.23861,−2.07153,−.1179,−.1157,−.0856,.89246,1.3566,1.4079, 1.763,1.8403,1.17878,−1.28207,−.14942,−0.14578,−.24326,−1.6167,−1.4481, −.45096,1.0881,.55007,.71771,.37682,−1.15096,−1.10557,−1.0576,−.69389,.6836, 1.1733,.5395,.07034,.73916,1.000165,−.08865,−.7606,−.09183,.71343,.3157.1.07053).

The likelihood function is

$$L(\theta|y) = (\tau/2\pi)^{n/2} \exp{-(\tau/2)\sum_{i=1}^{i=n}[y(t)-\theta y(t-1)]^2} \qquad (2.21)$$

where y is the n by 1 vector of observations. Suppose one assumes the prior density of the parameters is

$$\zeta(\theta, \tau) = 1/\tau, \tau > 0, -\infty < \theta < \infty, \tag{2.22}$$

Then, it can be shown that the posterior density of θ is

$$\zeta(\theta|y) \propto \left\{1 + \lambda(\theta - \nu)^2 1\right\}^{-n/2} \tag{2.23}$$

where

$$\nu = \sum_{t=2}^{t=n} y(t)y(t-1) / \sum_{t=2}^{t=n} y^2(t-1). \tag{2.24}$$

Thus, the posterior distribution of θ is a univariate t with $n - 1$ degrees of freedom, mean ν, and precision $\lambda = (n-1)/c$, where

$$c = \sum_{t=1}^{t=n} y^2(t) - [\sum_{t=2}^{n} y(t)y(t-1)]^2 / \sum_{t=2}^{t=n} y^2(t-1). \tag{2.25}$$

Using fifty observations of the vector y, it follows that
$c = 28.581, \lambda = 1.71442$, and $\nu = .6246$; thus, the posterior distribution of θ is a t with 49 degrees of freedom, mean .6246, and precision 1.7144.

This section will introduce the way MCMC techniques are used to execute the posterior analysis for the AR(1) time series model which was analyzed analytically above. Remember the vector y of observations from the process with parameters $\theta = .6$ and $\tau = 1$, and consider the WinBUGS code bellow.

BC 2.1
```
model;
v~dgamma(.01,.01)
#v is the precision τ with a noninformative gamma (.01,.01) distribution
theta~beta(6,4)
# the prior distribution of theta is beta(6,4)
# the vector Y is the vector of 50 observations
# Y follows a multivariate normal distribution with mean mu and variance
covariance matrix Sigma. The matrix tau is the inverse of the variance
covariance matrix
Y[1,1:50]~dmnorm(mu[],tau[,])f
for( i in 1:50){mu[i]<-0}
tau[1:50,1:50]<-inverse(Sigma[,])
for (i in 1:50){Sigma[i,i]<-v/(1-theta*theta)}
```

```
for( i in 1:50){ for j in i+1:50 (Sigma[i,j]<-v*pow(theta,j)*1/(1-theta*theta)}}
for( i in 2:50){for j in 1:i-1){Sigma[I,j]<-v*pow(theta,i-1)*1/(1-theta*theta)}}
}
# the following list statement is for the 50 by 1 observation vector
list(Y=structure (.Data=c(-.62,-.1922,-.9509,1.0247,.9443,-.2538,.3351,
.93938,1.13954,.37826,1.73873,1.43308,.23861,-2.07153,-.1179,-.1157,-.0856,
.89246,1.3566,1.4079,1.763,1.8403,1.17878,-1.28207,-.14942,-0.14578,-.24326,
-1.6167,-1.4481,-.45096,1.0881,.55007,.71771,.37682,-1.15096,-1.10557,
-1.0576,.-.69389,.6836,1.1733,.5395,.07034,.73916,1.000165,.-.08865,-.7606,
-.09183,.71343,.3157.1.07053),.Dim=c(1,50)))
# the following list statement specifies the initial values of MCMC process
list(theta=.6, v=1)
```

Note that the WinBUGS analysis did not assume the same prior as was assumed in the analytical approach; nevertheless, the two analyses should agree because the MCMC approach assumed a noninformative gamma prior. Also note that both analyses did not utilize the stationary restriction $|\theta| < 1$!

The Bayesian analysis is executed with 45,000 observations for the simulation with a 5,000 burn-in and a refresh of 100. The results are reported in Table 2.1.

The Bayesian analysis for the AR(1) model shows that the posterior mean of the correlation coefficient θ is .6258, which compares quite favorable to the value .6 used to generate the data. Note the posterior mean of .6258 is computed via WinBUGS, whereas the analytical value computed earlier is .6246, and the two values agree to the nearest 100th decimal.

Bayesian methods for autoregressive processes will be developed in more detail in Chapter 5.

2.4.3 The Poisson Distribution

The Poisson distribution often occurs as a population for a discrete random variable with mass function

$$f(X|\theta) = e^{-\theta}\theta^x/x!, \qquad (2.26)$$

TABLE 2.1

Posterior Analysis for AR(1) Series

Parameter	Mean	SD	Error	2 1/2	Median	97 1/2
θ	0.626	0.151	0.002	0.31	0.638	0.868
σ^2	0.572	0.214	0.003	0.228	0.552	1.043

where the gamma density

$$\xi(\theta) = [\beta^{\alpha}/\Gamma(\alpha)]\theta^{\alpha-1}e^{-\theta\beta}, \tag{2.27}$$

is a conjugate distribution that expresses informative prior information. For example, in a previous experiment with m observations, the prior density would be gamma with the appropriate values of alpha and beta. Based on a random sample of size n, the posterior density is

$$\xi(\theta|\text{data}) \propto \theta^{\sum_{i=1}^{i=n}x_i+\alpha-1}e^{-\theta(n+\beta)}, \tag{2.28}$$

which is identified as a gamma density with parameters $\alpha' = \sum_{i=1}^{i=n} x_i + \alpha$ and $\beta' = n + \beta$. Remember the posterior mean is α'/β', median $(\alpha' - 1)/\beta'$, and variance $\alpha'/(\beta')^2$.

One of the most important time series is the Poisson process. The Poisson process $N(t)$ with parameter $\lambda > 0$ is defined as follows:

1. $N(t)$ is the number of events occurring over time 0 to t with $N(0) = 0$ and the process has independent increments.

2. For all $t > 0, 0 \langle P[N(t)\rangle 0] < 1$, that is to say for all intervals, no matter how small, there is a positive probability that an event will occur, but it is not certain an event will occur.

3. For all $t \geq 0$,

$$\lim\{P[N(t+h) - N(t) \geq 2]/P[N(t+h) - N(t) = 1]\},$$

   where the limit is as h approaches 0. This implies that events cannot occur simultaneously.

4. The process has stationary independent increments; thus for all points $t > s \geq 0$ and $h > 0$, the two random variables $N(t+h) - N(s+h)$ and $N(t) - N(s)$ are identically distributed and are independent.

Based on these four axioms, one may show that for all $t > 0$, there exists a $\lambda > 0$ such that $N(t)$ has a Poisson distribution with mean λt. Thus, the average number of events occurring over $[0, t)$ is λt and the average number of events occurring per unit time is λ. The Poisson process is a counting process (it counts the number of events occurring over time) and has many generalizations that will be introduced in Chapter 7. An interesting feature of the Poisson process is that the time between the occurrence of two adjacent events has an exponential distribution. In particular, if $N(t), t \geq 0$ is a Poisson process with parameter λ, then the successive inter-arrival times are independent and have an exponential distribution with mean $1/\lambda$; thus, the Poisson process can be simulated via the exponential distribution. For example, consider a Poisson process with parameter $\lambda = 5$, and suppose a realization of 50 using the exponential distribution with mean $1/5 = .2$ is to be generated using the following WinBUGS code.

BC 2.2
```
model {
for (i in 1 : 1000) {
y[i] ~ dexp(.2)
}
}
```
The 50 successive inter arrival times are given by vector I.

I=(.403,11.00,23.11,1.92,.25,4.34,3.53,.10,1.59,.05,3.11,2.21,3.03,5.96,7.22,.96,
8.75,2.23,29.84,2.96,2.41,2.86,.5411.48,,2.10,1.43,8.99,6.87,1.73,6.76,14.91,11.90,
1.21,12.08,4.49,4.14,1.94,1.30,1.86,4.86,.21,13.27,.42,1.60,3.38,3.39,2.97,9.97,
7.03,2.54)

The fifty corresponding waiting times are the components of the vector W below:

W=(.40,11.40,34.51,36.43,36.68,41.02,44.55,44.65,46.24,46.29,49.40,51.61,54.64,
60.60,67.82,68.78,77.53,79.76,109.60,112.56,114.97,117.83,118.37,129.85,131.95,
133.38,142.37,149.24,150.97,157.73,172.64,184.54,185.75,197.83,202.32,206.46,
208.40,209.70,211.56,216.42,216.63,229.90,230.32,231.92,235.30,238.69,241.66,
251.63,258.66,261.20).

Thus, the first event occurred at time .403 time units and the second at 11.40 time units, and the last at 261.2 units.

Let T_n be the n-th inter-arrival time and W_n the corresponding waiting time, then

$$W_n = T_1 + T_2 + \cdots + T_n, \tag{2.29}$$

where $n = 0, 1, 2, \ldots$, thus, we know that

$$T_n \sim \exp(\lambda) \tag{2.30}$$

and

$$W_n \sim \mathrm{gamma}(n, \lambda). \tag{2.31}$$

That is the inter-arrival times have a common exponential distribution with parameter λ and n-th waiting time has a gamma distribution with parameters n and λ. In the next section on inference, based on the above inter-arrival and waiting times, Bayesian inferences for the intensity λ will be performed.

2.5 Inference

2.5.1 Introduction

In a statistical context, by inference, one usually means estimation of parameters, tests of hypotheses, and prediction of future observations. With the Bayesian approach, all inferences are based on the posterior distribution of the parameters, which in turn is based on the sample, via the likelihood function and the prior distribution. We have seen the role of the prior density and likelihood function in determining the posterior distribution, and presently will focus on the determination of point and interval estimation of the model parameters, and later will emphasize how the posterior distribution determines a test of hypothesis. Finally, the role of the predictive distribution in testing hypotheses and in goodness of fit will be explained.

When the model has only one parameter, one would estimate that parameter by listing its characteristics, such as the posterior mean, media, and standard deviation and plotting the posterior density. However, if there are several parameters, one would determine the marginal posterior distribution of the relevant parameters and, as above, calculate its characteristics (e.g., mean, median, mode, standard deviation, etc.) and plot the densities. Interval estimates of the parameters are also usually reported and are called credible intervals.

2.5.2 Estimation

Inferences for the normal (μ, τ) population are somewhat more demanding, because both parameters are unknown. Assuming the vague prior density $\xi(\mu, \tau) \propto 1/\tau$, the marginal posterior distribution of the population mean μ is a t-distribution with $n - 1$ degrees of freedom, mean $\bar{x}$, and precision n/s^2; thus, the mean and the median are the same and provide a natural estimator of μ, and because of the symmetry of the t-density, a $(1 - \alpha)$ credible interval for μ is $\bar{x} \pm t_{\alpha/2, n-1} S/\sqrt{n}$, where $t_{\alpha/2, n-1}$ is the upper 100 $\alpha/2$ percent point of the t-distribution with $n - 1$ degrees of freedom. To generate values from the $t(n - 1, \bar{x}, n/S^2)$ distribution, generate values from Student's t-distribution with $n - 1$ degrees of freedom, multiply each by $S/\sqrt{n}$, and then add $\bar{x}$ to each. Suppose $n - 30$,

x = (7.8902,4.8343,11.0677,8.7969,4.0391,4.0024,6.6494,8.4788,0.7939,5.0689,
6.9175,6.1092,8.2463,10.3179,1.8429,3.0789,2.8470,5.1471,6.3730,5.2907,1.5024,
3.8193,9.9831,6.2756,5.3620,5.3297,9.3105,6.5555,0.8189,0.4713), then

$\bar{x} = 5.57$ and $S = 2.92$.

Using the same dataset, the following WinBugs code is used to analyze
the problem.

BC 2.1

```
Model;
{ for( i in 1:30) { x[i]~dnorm(mu,tau) }
mu~dnorm (0.0,.0001)
tau ~dgamma( .0001,.0001)
sigma <- 1/tau }
list(x = c(7.8902,4.8343,11.0677,8.7969,4.0391,4.0024,6.6494,8.4788,0.7939, 5.0689,
6.9175,6.1092,8.2463,10.3179,1.8429,3.0789,2.8470,5.1471,6.3730,5.2907,1.5024,
3.8193,9.9831,6.2756,5.3620,5.3297,9.3105,6.5555,0.8189,0.4713))
```

list(mu = 0, tau = 1)

Note, that a somewhat different prior was employed here, compared to
previously, in that μ and τ are independent and assigned properly, but
noninformative distributions. The corresponding analysis gives the following as shown in Table 2.2.

Upper and lower refer to the lower and upper 2½ percent points of the
posterior distribution. Note a 95% credible interval for mu is (4.47, 6.65)
and the estimation error is .003566. See the Appendix for the details on
executing the WinBUGS statements above.

The program generated 30,000 samples from the joint posterior distribution of μ and σ using a Gibbs sampling algorithm, and used 29,000 for the
posterior moments and graphs, with a refresh of 100.

2.5.3 Testing Hypotheses

An important feature of inference is testing hypotheses. Often in stochastic
processes, the scientific hypothesis can be expressed in statistical terms and

TABLE 2.2

Posterior Distribution of μ and $\sigma = 1/\sqrt{\tau}$

Parameter	Mean	Std dev	MC error	Median	Lower	Upper
μ	5.572	0.5547	0.00357	5.571	4.479	6.656
σ	9.15	2.57	0.01589	8.733	5.359	15.37

a formal test implemented. Suppose $\Omega = \Omega_0 \cup \Omega_1$ is a partition of the parameter space, then the null hypothesis is designated as $H_0 : \theta \in \Omega_0$ and the alternative by $H_1 : \theta \in \Omega_1$, and a test of H_0 versus H_1 consists of rejecting H_0 in favor of H_1 if the observations $x = (x_1, x_2, \ldots, x_n)$ belong to a critical region C. In the usual approach, the critical region is based on the probabilities of type I errors, namely $\Pr(C|\theta)$, where $\theta \in \Omega_0$ and of type II errors $1 - \Pr(C|\theta)$, where $\theta \in \Omega_1$. This approach to testing hypothesis was developed formally by Neyman and Pearson and can be found in many of the standard references, such as Lehmann.[20] Lee[21] presents a good elementary introduction to testing and estimation in a Bayesian context.

In the Bayesian approach, the posterior probabilities

$$p_0 = \Pr(\theta \in \Omega_0 | data) \tag{2.32}$$

and

$$p_1 = \Pr(\theta \in \Omega_1 | data) \tag{2.33}$$

are required, and on the basis of the two, a decision is made whether or not to reject H in favor of A or to reject A in favor of H. Also required are the two corresponding prior probabilities

$$\pi_0 = \Pr(\theta \in \Omega_0) \tag{2.34}$$

and

$$\pi_1 = \Pr(\theta \in \Omega_1). \tag{2.35}$$

Now consider the prior odds π_0/π_1 and posterior odds p_0/p_1. In turn, consider the Bayes factor B in favor of H_0 relative to H_1, namely

$$B = (p_0/p_1)/(\pi_0/\pi_1), \tag{2.36}$$

Then, the posterior probabilities p_0 and p_1 can be expressed in terms of the Bayes factor, thus:

$$p_0 = 1/[1 + (\pi_1/\pi_1)B^{-1}] \tag{2.37}$$

and the Bayes factor is interpreted as the odds in favor of H_0 relative to H_1 as implied by the information from the data.

When the hypotheses are simple, that is, $\Omega_0 = \{\theta_0\}$ and $\Omega_1 = \{\theta_1\}$, note that the odds ratio can be expressed as the likelihood ratio.

$$B = p(x|\theta_0)/p(x|\theta_1). \tag{2.38}$$

This interpretation is not valid when Ω_0 and Ω_1 are composite. Consider the restriction of the prior density $p(\theta)$ to Ω_0, namely

$$p_0(\theta) = p(\theta)/\pi_0, \quad \theta \in \Omega_0 \tag{2.39}$$

and its restriction to Ω_1, namely

$$p_1(\theta) = p(\theta)/\pi_1. \quad \theta \in \Omega_1 \tag{2.40}$$

Note the integral of $p_0(\theta)$ with respect to θ over Ω_0 is 1.

Now it can be shown that that the posterior probability of the null hypothesis is

$$p_0 = \pi_0 \int p(x|\theta)p_0(\theta)d\theta \tag{2.41}$$

where the integral is taken over Ω_0.

In a similar way, the posterior probability of H_1

$$p_1 = \pi_1 \int p(x|\theta)p_1(\theta)d\theta \tag{2.42}$$

where the integral is taken over Ω_1.

Now the Bayes factor can be expressed as

$$B = (p_0/p_1)/(\pi_0/\pi_1)$$

$$= \int_{\theta \in \Omega_0} p(x|\theta)p_0(\theta)d\theta / \int_{\theta \in \Omega_1} p(x|\theta)p_1(\theta)d\theta \tag{2.43}$$

which is the ratio of weighted likelihood functions, weighted by the prior probability densities restricted to Ω_0 and Ω_1.

An important aspect of testing hypotheses is when the null hypothesis is a point null hypothesis and the alternative is composite; thus, consider

$$H_0 : \theta = \theta_0 \tag{2.44}$$

versus

$$H_1 : \theta \neq \theta_0 \tag{2.45}$$

where θ_0 is known. How does one assign prior information to this case? A reasonable approach is to assign a positive probability π_0 for the null hypothesis and for the alternative assign a prior density $\pi_1 p_1(\theta)$, where

$$\int_{\theta \neq \theta_0} p_1(\theta)d\theta = 1 \tag{2.46}$$

Thus, $\pi_0 + \pi_1 = 1$ and it is seen that the prior probability of the alternative is π_1 and for values $\theta \neq \theta_0$, p_1 is the density of a continuous random variable that expresses the prior knowledge one has for the alternative hypothesis.

Let

$$p(x) = \pi_0 p(x|\theta_0) + \pi_1 \int p_1(\theta)p(x|\theta)d\theta \tag{2.47}$$

where X is the vector of observations with conditional density $p(x|\theta)$ and where $p(x)$ is the marginal density of the observations.

By letting

$$p_1(x) = \int_{\theta \neq \theta_0} p_1(\theta)p(x|\theta)d\theta \tag{2.48}$$

The marginal density (2.58) can be expressed as

$$p(x) = \pi_0 p(x|\theta_0) + \pi_1 p_1(x) \tag{2.49}$$

and the posterior probabilities of the null and alternative hypotheses can be expressed as

$$p_0 = \pi_0 p(x|\theta_0)/[\pi_0 p(x|\theta_0) + \pi_1 p_1(x)]$$
$$= \pi_0 p(x|\theta_0)/p(x). \tag{2.50}$$

In a similar manner,

$$p_1 = \pi_1 p_1(x)/p(x) \tag{2.51}$$

for the posterior probability of the alternative hypothesis. If one desires to use the Bayes factor, then one may show

$$B = p(x|\theta_0)/p_1(x). \tag{2.52}$$

The above derivation of the posterior probabilities in the context of hypothesis testing closely follows Lee.[15]

In summary, for testing hypotheses via the Bayesian approach, the following is required:

1. The prior probabilities of the null and alternative hypotheses, namely π_0 and π_1.
2. The prior density $p_1(\theta)$ for values of $\{\theta : \theta \neq \theta_0\}$.
3. The likelihood function, that is the joint conditional density of the observations $x = (x_1, x_2, ..., x_n)$ given θ, for all values of θ in the parameter space.

For the first example in testing hypotheses when the null is simple but the alternative is composite, consider the AR(1) process

$$Y(t) = \theta Y(t-1) + W(t), t = 1, 2, \ldots \quad (2.15)$$

where
$$W(t), t = 1, 2, \ldots \quad (2.16)$$

is a sequence of independent normal random variables with mean zero and precision τ, and $\tau > 0$.

Consider the following testing problem using a Bayesian approach, where the null hypothesis is

$$H : \theta = .6 \quad (2.53)$$

versus the alternative
$$A : \theta \neq .6.$$

Recall that the null hypothesis value $\theta = .6$ is the value used to generate a sample of size $n = 50$ from the process; thus, we will be testing to see if the null hypothesis is indeed supported by the data generated from the model.

Recall that the likelihood function for θ and τ is

$$L(\theta, \tau | data) = (\tau/2\pi)^{n/2} \exp -(\tau/2)\left\{ \sum_{t=1}^{t=n} y^2(t-1)[\theta - \mu]^2 + c \right\}$$

where
$$\mu = \sum_{t=1}^{t=n} y(t)y(t-1) / \sum_{t=1}^{t=n} y^2(t-1)$$

and
$$c = \sum_{t=1}^{t=n} y^2(t) - [\sum_{t=1}^{t=n} y(t)y(t-1)]^2 / \sum_{t=1}^{t=n} y^2(t-1). \quad (2.54)$$

I use the marginal likelihood for θ

$$p(\theta|data) = \tau^{(n-1)/2}[(n-1)\pi]^{1/2}/(2\pi)^{n/2}c^{n/2}f(\theta|data) \qquad (2.55)$$

where

$$f(\theta|data) = \left\{\tau^{1/2}\Gamma(n/2)/[(n-1)\pi]^{1/2}\Gamma((n-1)/2)\right\}$$
$$\left\{1+(\tau/(n-1))[\theta-\mu]^2\right\}^{n/2}$$

and

$$\tau = (n-1)\sum_{t=1}^{t=n} y^2(t-1)/c. \qquad (2.56)$$

Note that $f(\theta|data)$ is the density of a t-distribution with $n-1$ degrees of freedom, location μ, and precision τ.

We now return to (2.47) and calculate the posterior probability p_0 of the null hypothesis.

Note that the posterior probability of the null hypothesis is

$$p_0 = [1 + \gamma_1 p_1(y)/\gamma_0 p(y|\theta_0)]^{-1} \qquad (2.57)$$

where γ_0 is the prior probability of the null hypothesis and $\gamma_1 = 1 - \gamma_0$. Also,

$$p(y|\theta) = \left\{\tau^{(n-1)/2}[(n-1)\pi]^{1/2}/(2\pi)^{n/2}c^{n/2}\right\}f(\theta|y) \qquad (2.54)$$

and $f(\theta|y)$ is the density of a univariate t-distribution with $n-1$ degrees of freedom, location μ, and precision τ.

Also,

$$p_1(y) = \int p_1(\theta)p(y|\theta)d\theta = \tau^{n/2}[(n-1)\pi]^{1/2}/(2\pi)^{n/2}c^{n/2}, \qquad (2.58)$$

thus, when the observations are inserted into (2.54), the posterior probability of the null hypothesis can be calculated. The student will be asked to calculate this probability based on the fifty observations generated from the AR(1) series (2.15).

An earlier and more informal approach[8] to testing hypotheses is to reject the null hypothesis if the 95% credible region for θ does not contain the set of all θ such that $\theta \in \Omega_0$. In the special case that $H:\theta = \theta_0$ versus the alternative $A: \theta \neq \theta_0$, where θ is a scalar, H is rejected when the 95% confidence interval for θ does not include θ_0. However, there are some logical problems with this approach. If a continuous prior density is used for the entire parameter space, the prior probability of the null hypothesis is zero, which implies a posterior probability of zero for the null hypothesis, thus, implying illogical reasoning for this approach to testing hypotheses!

In this book, hypothesis testing is an important component as does estimating the unknown parameters, however, therefore when testing called for the formal approach developed above in this section will be conducted.

2.6 Predictive Inference

2.6.1 Introduction

Our primary interest in the predictive distribution is to check for model assumptions. Is the adopted model for an analysis the most appropriate?

What is the predictive distribution of a future set of observations Z? It is the conditional distribution of Z given $X = x$, where x represents the past observations, which when expressed as a density is

$$g(z|X) = \int_{\Omega} f(z|\theta)\, \xi(\theta|X) d\theta,\ z \in R^m \qquad (2.59)$$

where the integral is with respect to θ, and $f = (x|\theta)$ is the density of $x = (x_1, x_2 \ldots, x_n)$, given θ. This assumes that given θ, that Z and X are independent. Thus, the predictive density is posterior average of $f = (z|\theta)$ with respect to the posterior distribution of θ.

The posterior predictive density will be derived for the binomial and normal populations.

2.6.2 The Binomial Population

Suppose the binomial case is again considered, where the posterior density of the binomial parameter θ is

$$\xi(\theta|X) = [\Gamma(\alpha+\beta)\Gamma(n+1)/\Gamma(\alpha)\Gamma(\beta)\Gamma(x+1)\Gamma(n-x+1)]$$
$$\theta^{\alpha+x-1}(1-\theta)^{\beta+n-x-1}, \qquad (2.60)$$

a beta with parameters $\alpha + x$ and $n - x + \beta$, and X is the sum of the set of n observations. The population mass function of a future observation Z is

$f(z/\theta) = \theta^z(1-\theta)^{1-z}$, and the predictive mass function of Z, called the beta-binomial, is

$$g(z|X) = \Gamma(\alpha+\beta)\Gamma(n+1)\Gamma(\alpha+\sum_{i=1}^{i=n}x_i+z)\Gamma(1+n+\beta-x-z)\div$$
$$\Gamma(\alpha)\Gamma(\beta)\Gamma(n-x+1)\Gamma(x+1)\Gamma(n+1+\alpha+\beta) \quad (2.61)$$

where $z=0.1$. Note this function does not depend on the unknown parameter, and that the n past observations are known, and that if $\alpha=\beta=1$, one is assuming a uniform prior density for θ.

As an example for the predictive distribution of the binomial distribution, let

$$N = \begin{pmatrix} 1,7,4,6,2 \\ 5,3,4,6,5 \\ 3,6,2,3,1 \\ 5,7,2,5,4 \\ 6,1,3,3,4 \end{pmatrix} \quad (2.62)$$

be the transition counts for a five-state Markov chain and

$$\Phi = \begin{pmatrix} \phi_{11},\phi_{12},\phi_{13},\phi_{14},\phi_{15} \\ \phi_{21},\phi_{22},\phi_{23},\phi_{24},\phi_{25} \\ \phi_{31},\phi_{32},\phi_{33},\phi_{34},\phi_{35} \\ \phi_{41},\phi_{42},\phi_{43},\phi_{44},\phi_{45} \\ \phi_{51},\phi_{52},\phi_{53},\phi_{54},\phi_{55} \end{pmatrix} \quad (2.63)$$

as the one-step transition matrix.

Our focus is on forecasting the number of transitions Z_{11} from 1 to 1, that is the number of times the chain remains in state 1, assuming a total of m replications for the first row of the chain, that is $Z_{11}=0,1,2,...,m$. Using (2.77), one may show the predictive mass function of Z_{11} is equation (2.78)

$$g(z|n_{11}=1) = \binom{m}{z}\Gamma(\alpha+\beta)\Gamma(n+1)\Gamma(\alpha+z+n_{11})\Gamma(\beta+n-z-n_{11})/$$
$$\Gamma(\alpha)\Gamma(\beta)\Gamma(n_{11}+1)\Gamma(n-n_{11}+1)\Gamma(\alpha+\beta+n) \quad (2.64)$$

The relevant quantities of (2.78) are $n=20$, $n_{11}=1$. Also remember that α and β are the parameters of the prior distribution of ϕ_{11}, the probability of remaining in state 1, and that predictive inferences are conditional on $n=20$, the total transition counts for the first row of the one-step transition matrix of the chain.

Another example of prediction is a binary time series, which is defined by

$$Y(t)|Y(t-1), Y(t-2), \ldots, Z(t-1), Z(t-2), \ldots \sim Bern(\pi_t(\gamma))$$
$$g(\pi_t(\gamma)) = \gamma Z(t-1) \qquad (2.65)$$
$$Z(t-1) = (Z_1(t-1), \ldots, Z_p(t-p))$$

where $Z(t-1)$ is a regressor which can be independent of the previous binary observations or can be only autoregressive, namely $Z(t-1) = (Y(t-1), Y(t-2), \ldots, Y(t-p))$. It can also be a mixture of both. See Wilks and Wilby[22] and also Chapter 10 of Davis et al.[23] for the article by Kirch and Kamgaing.[24]

Returning to the autoregressive binary time series (2.62), let the function g be the canonical link $g(x) = \log(x/(1-x))$, then Bayesian inferences can be based on the partial likelihood function

$$L(\gamma) = \prod_{t-1}^{t=n} \pi_t(\gamma)^{y(t)} (1 - \pi_t(\gamma))^{1-y(t)}, \qquad (2.66)$$

where γ is the 1 by p vector of regression parameters and the t-th binary observation is $y(t), t = 1, 2, \ldots, n$.

The example will be revisited in Chapter 15.

2.6.3 Forecasting from a Normal Population

Moving on to the normal density with both parameters unknown, what is the predictive density of Z, with noninformative prior density

$$\xi(\mu, \tau) = 1/\tau, \quad \mu \in R \text{ and } \tau > 0 \qquad (2.67)$$

The posterior density is

$$\xi(\mu, \tau|data) = [(\tau^{n/2-1}/(2\pi)^{n/2}] \exp -(\tau/2)[n(\mu - \bar{x})^2 + (n-1)S_x^2], \qquad (2.68)$$

where $\bar{x}$ and s_x^2 are the sample mean and variance, based on a random sample of size n, $x = (x_1. x_2, \ldots, x_n)$. Suppose z is a future sample $z = (z_1, z_2, \ldots, z_m)$ of size m, then the predictive density of Z is

$$g(z|X) = \int\int [\tau^{(n+m)/2-1}/(2\pi)^{(n+m)/2}] \exp -(\tau/2)[n(\mu - \bar{x})^2 \\ + (n-1)S_x^2 + m(\bar{z} - \mu)^2 + (m-1)S_z^2] \qquad (2.69)$$

where the integration is with respect to $\mu \in R$ and $\sigma > 0$.

m is the number of future observations, $\bar{z}$ is the sample mean of the m future observations, and S_z^2 is the corresponding sample variance.

It can be shown that the predictive density of z is

$$g(z|x) \propto \Gamma((n+m-1)/2)/[1+\zeta(\bar{z}-\bar{x})^2/(n+m-3)]^{(n+m-3+1)} \qquad (2.71)$$

where
$$\zeta = nm/(n+m)k(n+m-3) \qquad (2.72)$$

$$k = (n-1)S_x^2 + (m-1)S_z^2 + n^2(\bar{x})^2/(n+m). \qquad (2.73)$$

This density is recognized as a non-central t-distribution with $n+m-3$ degrees of freedom, location $\bar{x}$, and precision ζ.

The predictive distribution can be used as an inferential tool to test hypotheses about future observations, to estimate the mean of future observations, and to find confidence bands for future observations. In the context of stochastic processes, the predictive distribution for future normal observations will be employed to generate future values from a Wiener process.

Of interest in the context of time series is Brownian motion and the predictive distribution of the future observations z when $\mu = 0$, that is, when the posterior density is

$$g(\tau|x) \propto [\tau^{(n+m)/2-1}/(2\pi)^{(n+m)/2}] \exp(-(\tau/2)[\sum_{i-1}^{i=n} x_i^2 + \sum_{i=1}^{i=m} z_i^2]. \qquad (2.74)$$

This assumes the improper prior for τ,

$$\zeta(\tau) = 1/\tau, \tau > 0. \qquad (2.75)$$

For the Wiener process of Section 2.4, the process was sampled at times $t_1, t_2, \ldots, t_n$ with $t_1 < t_2 <, \ldots, < t_n$ and with independent increments

$$d_i = X_i - X_{i-1} \qquad (2.76)$$

and where
$$t_i - t_{i-1} = 1,$$

and $i = 1, 2 \ldots, n$.

Also,
$$t_0 = 0.$$

Consider m future increments

$$z_i = x_i - x_{i-1} \qquad (2.77)$$

for $i = n+1, \ldots, n+m$.
and time points satisfying

$$t_i - t_{i-1} = 1$$

Now assume the improper prior density (2.84) for τ, then the joint density of the d and z_i is

$$g(d, z|\tau) \propto \tau^{(n+m)/2-1} \exp(-\tau/2)[\sum_{i=1}^{i=n} d_i^2 + \sum_{i=n+1}^{i=n+m} z_i^2], \tau > 0. \qquad (2.78)$$

Thus, the predictive distribution of the m future independent increments is

$$g(z|d) \propto \Gamma((n+m)/2)/[1 + (\zeta/(n+m-1)) \sum_{i=n+1}^{i=n+m} z_i^2]^{(n+m-1+1)/2} \qquad (2.79)$$

This is recognized as a non-central t-density with n+m-1 degrees of freedom, location the zero vector, and precision $\xi = (n+m-1)/\sum_{i=1}^{i=n} d_i^2$.

Recall the Brownian motion example of Section 2.4.2 with a variance of .01 and where 100 observations from the process is designated by x. Using formula (2.79) with $m = 100$, $n = 100$, $\sum_{i=1}^{i=100} d_i^2 = 1.6458374, \zeta = 120.64514$, then based on **BUBS Code 2.3,** with the 100 predicted z values appear as the vector z.

BC 2.3

```
model {
for (i in 1 : 100) {
y[i] ~ dt(0, 120.645, 199)
}
}
```

z = c(−0.05473,−0.119,0.1056,−0.129,0.05168,−0.06759,0.07575,0.04806,0.003992,
−0.1113,−0.154,−0.009263,0.09279,−0.06837,−0.07757,−0.1289,0.03088,

0.09818,0.01693,−0.04028,−0.1602,0.09864,−0.05848,0.002767,−0.1908,−0.1578,
−0.004863,0.04017,−0.05318,0.08215,−0.0231,0.1652,0.01179,0.151,−0.2395,
0.00945,−0.05023,−0.09512,−0.04164,0.09382,−0.01882,−0.1193,0.03329,
0.02761,−0.07163,−0.05162,0.04595,0.108,0.01209,0.09053,−0.08401,0.08781,
−0.05834,−0.09858,0.1072,0.1007,0.04107,0.222,0.1023,−0.003405,−0.002853,
0.1584,0.05611,0.05067,0.04823,0.02001,0.1747,−0.1451,−0.0137,−0.1187,
0.04217,−0.01667,−0.04725,0.00841,0.09915,−0.05576,0.02669,0.04407,0.03509,
0.06624,0.05622,−0.05857,−0.1255,−0.03296,−0.128,−0.0193,−0.05927,−0.1122,
0.06573,0.06395,0.044,0.04435,0.04717,−0.1504,0.06941,−0.03644,−0.04695,
−0.1194,−0.003718,−0.08247))

To see the accuracy of the above predicted values, the sample mean and variance should be computed. How close to zero is the sample mean and how close to .01 is the sample variance of the Brownian motion example?

The next example for predicting future observations is based on the AR (p) series, which is given by

$$\theta(B)Y(t) = w(t), t = 1, 2, 3, \ldots \tag{2.80}$$

where

$$\theta(B) = 1 - \theta_1 B - \theta_2 B^2 - \cdots - \theta_p B^p \tag{2.81}$$

and the backshift operator is

$$B^i Y(t) = Y(t - i), i = 1, 2, \ldots \tag{2.82}$$

The series is stationary if the roots of $\theta(B) = 0$ lie outside the unit circle in the complex plain.

Let $Y(t), t = 1, 2, \ldots, n$ be n observations from the AR(p) series (2.77), where $n > p$, and $Y(0), Y(-1), Y(-2), \ldots, Y(1 - p)$ are the initial observations, and the $W(t), t = 1, 2, \ldots, n$ are i.i.d. normal random variables with mean 0 and precision τ.

Note that the joint density of the observations is

$$f(y|\theta, \tau) = (\tau/2\pi)^{n/2} \exp -(\tau/2) \sum_{t=1}^{t=n} [Y(t) - \theta_1 Y(t - 1) - \cdots - \theta_p Y(t - p)]^2$$

where the $\theta_i, i = 1, 2, \ldots, p$ are real unknown autoregressive parameters; then, based on this likelihood function, the predictive density of future observations will be derived.

One must specify the prior distribution of the vector of autoregressive parameters and the precision; thus, let the conditional prior of θ given τ be the multivariate normal density

$$\zeta(\theta|\tau) \propto \tau^{p/2} \exp -(\tau/2)(\theta - \mu)'P(\theta - \mu) \tag{2.83}$$

where θ is the p $\times$ 1 vector of autoregressive parameters and P is a $p \times p$ known precision matrix.

Also, the marginal prior density of τ is

$$\zeta(\tau) \propto \tau^{a-1} \exp -\tau\beta, \ \tau > 0 \tag{2.84}$$

a gamma density with known parameters α and β

It should be brought to the reader's attention that the noninformative improper density

$$\zeta(\tau) = 1/\tau, \tau > 0 \tag{2.85}$$

is appropriate when there is very little information.

Combining the likelihood function with the prior densities given by (2.80) and (2.81), the posterior density of θ and τ is

$$\zeta(\theta, \tau|y) \propto \tau^{(n+2a+p)/2} \exp -(\tau/2)[(\theta - A^{-1}C)'A(\theta - A^{-1}C) + D] \tag{2.83}$$

where

$$A = P + (a_{ij}), \tag{2.86}$$

$$C = P\mu + (c_j), \tag{2.87}$$

and

$$D = 2\beta + \mu'P\mu + \sum_{t=1}^{t=n} Y^2(t) - C'A^{-1}C, \tag{2.88}$$

Also, (a_{ij}) is a matrix of order p with ij-th element

$$a_{ij} = \sum_{t=1}^{t=n} Y(t-i)Y(t-j) \tag{2.89}$$

and the p vector (c_j) has j-th component

$$c_j = \sum_{t=1}^{t=n} Y(t)Y(t-j). \tag{2.90}$$

Since the joint prior density of θ and τ given by (2.80) and (2.81) is known as a normal-gamma and is conjugate for this population, namely the posterior distribution of θ and τ will also be normal-gamma.

Now integrating the joint density (2.83) with respect to τ gives the marginal posterior density of θ as

$$\zeta(\theta|y) \propto [(\theta - A^{-1}C)'A(\theta - A^{-1}C) + D]^{-(n+2\alpha+p)/2}, \theta \in R^p \qquad (2.91)$$

Thus, the marginal posterior distribution of θ is a multivariate t with $n + 2\alpha$ degrees of freedom, location vector $A^{-1}C$, and precision matrix $(n + 2\alpha)A^{-1}D$. In addition, the marginal posterior density of τ is

$$\zeta(\tau|data) \propto \tau^{(n+2\alpha)/2-1} \exp(\tau/2)[2\beta + \mu'P\mu + \sum_{t=1}^{t=n} Y^2(t) - CA^{-1}C] \qquad (2.92)$$

Therefore, the posterior distribution of τ is gamma with parameters $(n + 2\alpha)/2$ and $[2\beta + \mu'P\mu + \sum_{t=1}^{t=n} Y^2(t) - C'A^{-1}C]/2$.

Consider one future observation $Y(n + 1)$; then, its conditional density given θ, τ, y is

$$f(y(n + 1)|\theta, \tau, y) \propto \tau^{1/2} \exp -(\tau/2)[y(n + 1) - \theta_1 y(n) - \ldots - \theta_p y(n - (p - 1))]^2$$

This density can also be expressed as

$$f(y(n + 1)|\theta, \tau, y) \propto \tau^{1/2} \exp -(\tau/2)[y^2(n + 1) + \theta'E\theta - 2\theta'F] \qquad (2.93)$$

where E is of order $p+1$ with ij-th element $e_{ij} = y(n - (i - 1))y(n - (j - 1))$,

and F is a $p+1$ vector with j-th element $y(n + 1)y(n - (j + 1))$. Now multiplying the posterior density (2.83) with the conditional predictive density (2.90) gives for the resulting exponent a perfect square in θ; now integrating this product with respect to θ and τ results in the predictive density

$$g(y(n + 1)|y) \propto [y^2(n + 1) + D + C'A^{-1}C - G]^{-(n+2\alpha+1)/2}, \qquad (2.94)$$

where

$$G = (C + F)'(A + E)^{-1}(C + F).$$

The predictive density can be simplified to a univariate t density with $(n + 2\alpha)$ degrees of freedom, location,

$$E[y(n+1)] = [1 - F^{*'}(A+E)^{-1}F]^{-1}F^{*'}(A+E)^{-1}C,$$

and precision
$$P[y(n+1)] = [1 - F^{*'}(A+E)^{-1}F^*]H, \tag{2.95}$$

where

$$H = D + C'A^{-1}C - C'(A+E)^{-1}F^*[1 - F^{*'}(A+E)^{-1}F^*]^{-1}F^{*'}(A+E)^{-1}C.$$

The p × 1 vector F^* has j-th component $y(n-(j-1))$, $j = 1, 2, \ldots, p$. For additional information about the predictive distribution, see Chapter 5 of Broemeling.[12]

Of course, the predictive density of say m future observations can be derived in the same manner as above.

2.7 Checking Model Assumptions

2.7.1 Introduction

It is imperative to check model adequacy in order to choose an appropriate model and to conduct a valid study. The approach taken here is based on many sources, including Gelman et al.,[14] Carlin and Louis,[18] and Congdon[16] Our main focus will be on the likelihood function of the posterior distribution, and not the prior distribution, and to this end, graphical representations such as histograms, boxplots, and various probability plots of the original observations will be compared with those of the observations generated from the predictive distribution. In addition to graphical methods, Bayesian versions of overall goodness-of-fit-type operations are taken to check model validity. Methods presented at this juncture are just a small subset of those presented in more advanced works, including Gelman et al., Carlin and Louis, and Congdon.

Of course, the prior distribution is an important component of the analysis, and if one is not sure of the 'true' prior, one should perform a sensitivity analysis to determine the robustness of posterior inferences to various alternative choices of prior information. See Gelman et al. or Carlin and Louis for details of performing a sensitivity study for prior information. Our approach is to either use informative or vague prior distributions, where the former is done when prior relevant experimental evidence determines the prior, or the latter is taken if there is none or very little germane experimental studies. In scientific studies, the most likely scenario is that there are relevant experimental studies providing informative prior information.

2.7.2 Sampling from an Exponential, but Assuming a Normal Population

Consider a random sample of size 30 from an exponential distribution with mean 3. An exponential distribution is often used to model the survival times of a screening test. The 30 exponential values are generated with BC 2.4.

BC 2.4

model;

```
{ for ( i in 1:30){
X[i]~dexp(3)
}}
```

x = (1.9075,0.7683,5.8364,3.0821,0.0276,15.0444,2.3591,14.9290,6.3841,7.6572,
5.9606,1.5316,3.1619,1.5236,2.5458,1.6693,4.2076,6.7704,7.0414,1.0895,3.7661,
0.0673,1.3952,2.8778,5.8272,1.5335,7.2606,3.1171,4.2783,0.2930).

The sample mean and standard deviation are 4.13 and 3.739, respectively.

Assume the sample is from a normal population with unknown mean and variance, with an improper prior density $\xi(\mu,\tau) = 1/\tau$, $\mu \in R$ and $\sigma > 0$, then the posterior predictive density is a univariate t with $n - 1 = 29$ degrees of freedom, mean $\bar{x} = 3.744$, standard deviation 3.872, and precision $p = .0645$. This is verified from the original observations x and the formula for the precision. From the predictive distribution, 30 observations are generated with BC 2.5.

BC 2.5.

```
{
For ( t in 1:30){
Z[t]~t(3.744,.0645,29)}
}
```

z = (2.76213,3.46370,2.88747,3.13581,4.50398,5.09963,4.39670,3.24032,3.58791,
5.60893,3.76411,3.15034,4.15961,2.83306,3.64620,3.48478,2.24699,2.44810,
3.39590,3.56703,4.04226,4.00720,4.33006,3.44320,5.03451,2.07679,2.30578,
5.99297,3.88463,2.52737)

which gives a mean of $\bar{z} = 3.634$ and standard deviation S =.975.

The histograms obviously are different, where for the original observations, a right skewness is depicted; however, this is lacking for the histogram of the predicted observations, which is for a t-distribution. Although the example seems trivial, it would not be for the first time that exponential observations were analyzed as if they were generated from a normal population! Of course, we have seen the relevance of the exponential distribution to Markov jump processes such as the Poisson process of Section 3.4.3, where the inter-arrival times are independent and identically distributed with a mean that is the reciprocal of the mean rate of event happenings.

It would be interesting to generate more replicate samples from the predictive distribution in order to see if these conclusions hold firm.

2.7.3 A Poisson Population

It is assumed the sample is from a Poisson population with mean 5; however, actually, it is generated from a uniform discrete population over the integers from 0 to 10. The sample of size 25 is

$x = (8,3,8,2,6,1,0,2,4,10,7,9,5,4,8,4,0,9,0,3,7,10,7,5,1)$, with a sample mean of 4.92 and standard deviation 3.278. When the population is Poisson, $P(\theta)$, and an uninformative prior

$$\xi(\theta) = 1/\theta, \theta > 0$$

is appropriate, the posterior density is gamma with parameters $alpha = \sum_{i=1}^{i=25} x_i = 123$ and beta = $n = 25$. Observations z from the predictive distribution are generated by taking a sample θ from the gamma posterior density, then selecting a z from the Poisson distribution $P(\theta)$. This was repeated 25 times to give

$$z = (2,5,6,2,4,3,5,3,2,3,3,6,7,5,5,3,1,5,7,3,5,3,6,4,5),$$

with a sample mean of 4.48 and standard deviation 1.896.

The most obvious difference shows a symmetric sample from the discrete uniform population, but, on the other hand, box plots of the predicted observations reveal a slight skewness to the right. The largest difference is in the inter-quartile ranges being (2, 8) for the original observations and (3, 5.5) for the predictive sample. Although there are some differences, to declare that the Poisson assumption is not valid might be premature. Of course, to reiterate, the Poisson process is one of the most important Markov time series.

2.7.4 The Weiner Process

Consider an example of a normal time series with $n = 100$ observations was simulated with R from a Wiener process with $\sigma = .01$ as the parameter.

Figure 2.1 depicts the values simulated from the Wiener process with observations taken 100 times one unit of time apart.

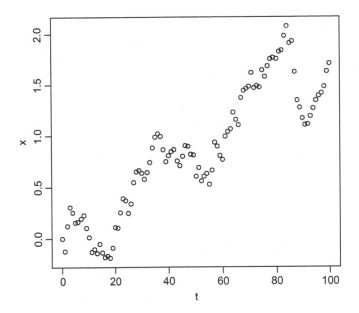

FIGURE 2.1
Simulation of the Wiener Process

R code 2.3 below is used to generate the observations and to plot the values over time.

RC 2.3.

```
t<-0:100
sig2<-.01
x<-rnorm(n=length(t)-1,sd=sqrt(sig2))
x<-c(0,cumsum(x))
plot(t,x,type="1",ylim=c(-2,2))
```

The vector d below is the corresponding 100 increment values with mean 0 and variance .01.

d=(−.1206600,2432790,.1833210,−.0540060,−.0933370,.0013660,.0350202,
.0358970,−.1266960,−.0904420,−.1401550,.0230410,−.0385080,.0894620,
−.0839162,−.0457920,.3528350,−.3555310,.0930150,.2003558,.0044500,
.1478410,.1384490,.0184600,.1224990,.0912220,.2035300,.1056839,.0123930,
−.0283500,−.0561130,.0625630,.1029310,.1417030,.0999980,.0343503,
−.0225143,−.1338220,−.1186080,.0298170,.0377400,.0186340,−.1070220,
−.0477170,.0933260,.1006602,−.0022484,−.0793380,−.0035692,−.2114558,
.0875840,−.1339130,.0489390,.0227763,−.1003730,.1335010,.2732970,
−.0419390,−.0883350,−.0381000,.2297750,.4127186,.0145250,.1665200,

−.0748080,−.0491901,.2643080,,.0693873,.0212337,.0179080,.1386532,
−.1509560,.0212460,−.0165320,.1709280,−.0679698,.1038280,.0764160,
.0083440,−07920,.0746620,.0122130,.1434140,.0934870,−.1668680,.0170720,
−.3026848,−.2791680,−.0691200,−.1085340,−.0616370,.0048761,.0871080,
.0826430,.0771660,.0452680,.0269480,.0646110,.1461450,.0801460)

These 100 observations should be from a normal population with mean 0 and variance .01. What does the histogram of these values indicate about the normal assumption? See Figure 2.2.

The sample mean is $\bar{x} = .0204995$ and the sample standard deviation is $s = .1237805$ or a sample variance of $s^2 = .01622$. These values are fairly close to the values 0 and .01, respectively, a good indication that the normality distribution of the increments is a reasonable assumption.

2.7.5 Testing the Multinomial Assumption

Consider a multinomial distribution where n_i is the number of times the state i occurs, where $i = 0, 1, 2, 3, 4$, and let p_i be the corresponding probability that state i occurs, then the probability mass function for the multinomial is

$$f(n_1, n_2, n_3, n_4, n_5 | p_1, p_2, p_3, p_4, p_5) \propto \prod_{i=0}^{i=4} p_i^{n_i} \tag{2.96}$$

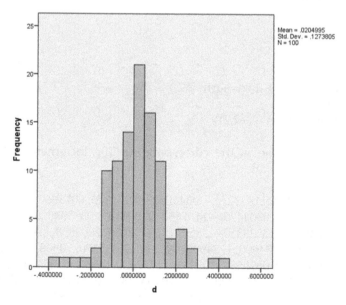

FIGURE 2.2
Histogram of the 100 Simulated Values from a Wiener Process

where $\sum_{i=0}^{i=4} p_i = 1$ and $\sum_{i=0}^{i=4} n_i = n$ with n known and fixed. The vector

$$n = \begin{pmatrix} n_1 \\ n_2 \\ n_3 \\ n_4 \\ n_5 \end{pmatrix} \tag{2.97}$$

is said to have a multinomial distribution with parameters n and

$$p = \begin{pmatrix} p_1 \\ p_2 \\ p_3 \\ p_4 \\ p_5 \end{pmatrix}. \tag{2.98}$$

Note that the constraint $\sum_{i=0}^{i=4} n_i = n$ implies that the components of n (2.92) are correlated, and it can be shown that

$$\text{cov}(n_i, n_j) = -np_i p_j. \tag{2.99}$$

We have seen that the multinomial distribution can be used to generate values from a Markov chain. Thus, consider the following R code that allows one to generate observations from a multinomial:

RC 2.4

$$\begin{aligned} & n < -\ 1000 \\ & \text{prob} < -\ c(.2,.2,.2,.2,.2) \\ & \text{rmultinom}(1, n, \text{prob}) \end{aligned} \tag{2.100}$$

$n = 1000$ is the sample size and prob the vector of multinomial probabilities .2,.2,.2,.2,.2 referring to the probabilities of the five categories.
Thus, consider the output of the execution of the R code (2.95):

$$\begin{aligned} & [0]\ 215 \\ & [1]\ 174 \\ & [2]\ 205 \\ & [3]\ 204 \\ & [4]\ 202 \end{aligned} \tag{101}$$

The simulation gives $n_0 = 215, n_1 = 174, n_2 = 205, n_3 = 204$, and $n_4 = 202$, that is, the number of zeros is 214, the number of ones 174, the number of twos 205, the number of threes 204, and the number of fours 202. Based on the multinomial probability vector prob = (.2,.2,.2,.2,.2), one would expect to have 200 outcomes for each of the five categories.

Our concern is: what is the accuracy of the R code in generating the five possible values, that is, does the above realization (2.96) represent a sample from the multinomial distribution?

Note that one could use the chi-square goodness of fit test to test the hypothesis

$$H{:}p_i = .2 \tag{2.102}$$

versus the alternative

$$A{:}p_i \neq .2$$

for i = 0,1,2,3,4.

The test statistic is

$$\chi^2 = \sum_{i=0}^{i=4} (n_i - 200)^2/200 = 1.125 + 3.38 + .125 + .08 + .02 = 4.91$$

which when compared to the five percentile of the chi square distribution with 4 degrees of freedom, $\chi^2_{4,.05} = .71$, implies the simulated values (2.96) were indeed generated from a multinomial distribution with probabilities given by (2.97). There is not enough evidence to reject the null hypothesis. Of course, a Bayesian approach to testing hypotheses is preferred; see Lee[13] for a detailed account for this approach. The student will be asked to use Bayesian techniques to test the above null hypothesis that $p_i = .2$.

Thus, in the case of time series, the best way to determine validity of the multinomial model is to know the details of how the study was designed and conducted. It is often the case that the details of the study are not available. The other important aspect of a multinomial population is that the probability of a particular outcome is constant over all cases. One other statistical way to check is to look for runs in the sequence, etc.

2.8 Computing

2.8.1 Introduction

This section introduces the computing algorithms and software that will be used for the Bayesian analysis of problems encountered in agreement investigations. In the previous sections of the chapter, direct methods (non-iterative) of computing the characteristics of the posterior distribution were demonstrated with some standard one-sample and two-sample problems. An example of this is the posterior analysis of a normal population, where the posterior distribution of the mean and variance is generated from its posterior distribution by the *t*-distribution random number generator in Minitab. In addition to some direct methods, iterative algorithms are briefly explained.

MCMC methods (an iterative procedure) of generating samples from the posterior distribution are introduced, where the Metropolis–Hasting algorithm and Gibb sampling are explained and illustrated with many examples. WinBUGS uses MCMC methods such as the Metropolis–Hasting and Gibbs sampling techniques, and many examples of a Bayesian analysis are given. An analysis consists of graphical displays of various plots of the posterior density of the parameters, by portraying the posterior analysis with tables that list the posterior mean, standard deviation, median, and lower and upper 2½ percentiles, and of other graphics that monitor the convergence of the generated observations.

2.8.2 Monte–Carlo–Markov Chain

2.8.2.1 *Introduction*

MCMC techniques are especially useful when analyzing data with complex statistical models. For example, when considering a hierarchical model with many levels of parameters, it is more efficient to use a MCMC technique such as Metropolis–Hasting or Gibbs sampling iterative procedure in order to sample from the many posterior distributions. It is very difficult, if not impossible, to use non-iterative direct methods for complex models.

A way to draw samples from a target posterior density $\xi(\theta|x)$ is to use Markov chain techniques, where each sample only depends on the last sample drawn. Starting with an approximate target density, the approximations are improved with each step of the sequential procedure. Or in other words, the sequence of samples is converging to samples drawn at random from the target distribution. A random walk from a Markov chain is simulated, where the stationary distribution of the chain is the target density, and the simulated values converge to the stationary distribution or the target density. The main concept in a Markov chain simulation is to devise a Markov process whose stationary distribution is the target density. The simulation must be long enough so that the present samples are close enough to the target. It has been shown that this is possible and that convergence can be accomplished. The general scheme for a Markov chain simulation is to create a sequence θ_t, $t=1,2,\ldots$ by beginning at some value θ_0, and at the *t*-th stage, select the present value from a transition function

$Q_t(\theta_t|\theta_{t-1})$, where the present value θ_t only depends on the previous one, via the transition function. The value of the starting value θ_0 is usually based on a good approximation to the target density. In order to converge to the target distribution, the transition function must be selected with care. The account given here is a summary of Gelman et al.,[19, ch. 11] who presents a very complete account of MCMC. Metropolis–Hasting is the general name given to methods of choosing appropriate transition functions, and two special cases of this are the Metropolis algorithm and the other is referred to as Gibbs sampling.

2.8.2.2 *The Metropolis Algorithm*

Suppose the target density $\xi(\theta|x)$ can be computed, then the Metropolis technique generates a sequence θ_t, $t=1,2,\ldots$ with a distribution that converges to a stationary distribution of the chain. Briefly, the steps taken to construct the sequence are:

    a. Draw the initial value θ_0 from some approximation to the target density,

    b. For $t = 1,2,\ldots$ generate a sample θ_* from the jumping distribution $G_t(\theta_*|\theta_{t-1})$,

    c. Calculate the ratio s $=\xi(\theta_*|X)\xi(\theta_{t-1}|X)$ and

    d. Let $\theta_t = \theta_*$ with probability min(s,1) or let $\theta_t = \theta_{t-1}$.

To summarize the above, if the jump given by b above increases the posterior density, let $\theta_t = \theta_*$; however, if the jump decreases the posterior density, let $\theta_t = \theta_*$ with probability s, otherwise let $\theta_t = \theta_{t-1}$. One must show the sequence generated is a Markov chain with a unique stationary density that converges to the target distribution. For more information, see Gelman et al.[19, p. 325]

2.8.2.3 *Gibbs Sampling*

Another MCMC algorithm is Gibbs sampling that is quite useful for multi-dimensional problems and is an alternating conditional sampling way to generate samples from the joint posterior distribution. Gibbs sampling can be thought of as a practical way to implement the fact that the joint distribution of two random variables is determined by the two conditional distributions.

The two-variable case is first considered by starting with a pair (θ_1, θ_2) of random variables. The Gibbs sampler generates a random sample from the joint distribution of θ_1 and θ_2 by sampling from the conditional distributions of θ_1 given θ_2 and from θ_2 given θ_1. The Gibbs sequence of size k

$$\theta_2^0, \theta_1^0; \theta_2^1, \theta_1^1; \theta_2^2, \theta_1^2; ...; \theta_2^k, \theta_1^k \qquad (2.103)$$

is generated by first choosing the initial values θ_2^0, θ_1^0 while the remaining are obtained iteratively by alternating values from the two conditional distributions. Under quite general conditions, for large enough k, the final two values θ_2^k, θ_1^k are samples from their respective marginal distributions. To generate a random sample of size n from the joint posterior distribution, generate the above Gibbs sequence n times. Having generated values from the marginal distributions with large k and n, the sample mean and variance will converge to the corresponding mean and variance of the posterior distribution of (θ_1, θ_2).

Gibbs sampling is an example of an MCMC because the generated samples are drawn from the limiting distribution of a 2 by 2 Markov chain. See Casella and George[25] for a proof that the generated values are indeed values from the appropriate marginal distributions. Of course, Gibbs sequences can be generated from the joint distribution of three, four, and more random variables.

The Gibbs sampling scheme is illustrated with a case of three random variables for the common mean of two normal populations.

2.8.2.4 The Common Mean of Normal Populations

Gregurich and Broemeling[26] describe the various steps in Gibbs sampling to determine the posterior distribution of the parameters in independent normal populations with a common mean.

The Gibbs sampling approach can best be explained by illustrating the procedure using two normal populations with a common mean θ. Thus, let $y_{ij}, j = 1, 2, ..., n_i$ be a random sample of size n_i from a normal population for $i = 1, 2$.

The likelihood function for θ, τ_1, and τ_2 is

$$L(\theta, \tau_1, \tau_2 | \text{data}) \propto$$

$$\tau_1^{\frac{n_1}{2}} \exp -\frac{\tau_1}{2} \left[(n_1 - 1)s_1^2 + n_1 (\theta - \bar{y}_1)^2 \right] * \tau_2^{\frac{n_2}{2}} \exp -\frac{\tau_2}{2} \left[(n_2 - 1)s_2^2 + n_2 (\theta - \bar{y}_2)^2 \right],$$

where $\theta \in \Re$, $\tau_1 > 0$, $\tau_2 > 0$, $s_1^2 = \sum_{j=1}^{n_1} \left(y_{1j} - \bar{y}_1 \right)^2 / (n_1 - 1)$, and

$$s_2^2 = \sum_{j=1}^{n_2} \left(y_{2j} - \bar{y}_2 \right)^2 / (n_2 - 1).$$

The prior distribution for the parameters θ, τ_1, and τ_2 is assumed to be a vague prior defined as

$$g(\theta, \tau_1, \tau_2) \propto \frac{1}{\tau_1}\frac{1}{\tau_2}, \quad \tau_i > 0,$$

Then, combining the above gives the posterior density of the parameters as

$$p(\theta, \tau_1, \tau_2 | \text{data}) \propto \prod_{i=1}^{2} \tau_i^{\frac{n_i-1}{2}} Exp - \frac{\tau_i}{2}\left[(n_i - 1)s_i^2 + n_i(\theta - \bar{y}_i)^2\right].$$

Therefore, the conditional posterior distribution of τ_1 and τ_2 given θ is such that

$$\tau_i | \theta \sim Gamma\left[\frac{n_i}{2}, \frac{(n_i - 1)s_i^2 + n_i(\theta - \bar{y}_i)^2}{2}\right], \qquad (2.104)$$

for $i = 1,2$ and given θ, τ_1, and τ_2 are independent.

The conditional posterior distribution of θ given τ_1 and τ_2 is normal. It can be shown that

$$\theta | \tau_1, \tau_2 \sim N\left[\frac{n_1\tau_1\bar{y}_1 + n_2\tau_2\bar{y}_2}{n_1\tau_1 + n_2\tau_2}, (n_1\tau_1 + n_2\tau_2)^{-1}\right]. \qquad (2.105)$$

Given the starting values $\tau_1^{(0)}$, $\tau_2^{(0)}$, and $\theta^{(0)}$ where

$$\tau_1^{(0)} = 1/s_1^2, \quad \tau_2^{(0)} = 1/s_2^2, \text{ and } \theta^{(0)} = \frac{n_1\bar{y}_1 + n_2\bar{y}_2}{n_1 + n_2},$$

draw $\theta^{(1)}$ from the normal conditional distribution of θ, given $\tau_1 = \tau_1^{(0)}$ and $\tau_2 = \tau_2^{(0)}$. Then, draw $\tau_1^{(1)}$ from the conditional gamma distribution (43), given $\theta = \theta^{(1)}$. And finally draw $\tau_2^{(1)}$ from the conditional gamma distribution of τ_2 given $\theta = \theta^{(1)}$. Then, generate

$$\theta^{(2)} \sim \theta | \tau_1 = \tau_1^{(1)}, \tau_2 = \tau_2^1$$

$$\tau_1^{(2)} \sim \tau_1 | \theta = \theta^2$$

$$\tau_2^{(2)} \sim \tau_2 | \theta = \theta^2.$$

Continue this process until there are t iterations $\left(\theta^{(t)},\ \tau_1^{(t)},\ \tau_2^{(t)}\right)$. For large t, $\theta^{(t)}$ would be one sample from the marginal distribution of θ, $\tau_1^{(t)}$ from the marginal distribution of τ_1, and $\tau_2^{(t)}$ from the marginal distribution of τ_2.

Independently repeating the above Gibbs process m times produces m 3-tuple parameter values $\left(\theta_j^{(t)},\ \tau_{1j}^{(t)},\ \tau_{2j}^{(t)}\right), j = 1, 2, \ldots, m$ which represent a random sample of size m from the joint posterior distribution of (θ, τ_1, τ_2). The statistical inferences are drawn from the m sample values generated by the Gibbs sampler.

The statistical inferences can be drawn from the m sample values generated by the Gibbs sampler. The Gibbs sampler will produce three columns, where each row is a sample drawn from the posterior distribution of (θ, τ_1, τ_2). The first column is the sequence of the sample m, the second column is a random sample of size m from the poly-t-distribution of θ, the third and fourth columns are also random samples of size m but from the marginal posterior distributions of τ_1 and τ_2, respectively.

Additional characteristics such as the median, mode, and the 95% credible region of the posterior distribution of the parameter θ can be calculated from the samples generated by the Gibbs technique. Hypothesis testing can also be performed. Similar characteristics of the parameters $\tau_1, \tau_2, \ldots, \tau_k$ can be calculated from the samples resulting from the Gibbs method.

2.8.2.5 An Example

The example is from Box and Tiao.[10, p. 481] It is referred to as 'the weighted mean problem.' It has two sets of normally distributed independent samples with a common mean and different variances. Samples from the posterior distributions were generated from Gibbs sequences using the statistical software Minitab®. The final value of each sequence was used to approximate the marginal posterior distribution of the parameters $\theta, \tau_1, \ldots, \tau_k$. All Gibbs sequences were generated holding the value of t equal to 50. Each example has the results of the parameters using four different Gibbs sampler sizes, where the sample size m is equal to 250, 500, 750, and 1500.

The 'weighted mean problem' has two sets of normally distributed independent observations with a common mean and different variances. The estimated posterior moments of θ determined by the Gibbs sampling method are reported in Table 2.3. The mean value of the posterior distribution of θ generated from the 250 Gibbs sequences is 108.42 with 0.07 as the standard error of the mean. The mean value of θ generated from 500 and 750 Gibbs sequences has the same value of 108.31, and the standard errors of the mean equal 0.04 and 0.03, respectively. The mean value of θ generated from 1500 Gibbs sequences is 108.36 and a standard error of the mean of 0.02. Box and Tiao determined the posterior distribution of θ using the t-distribution as an approximation to the

TABLE 2.3

Results from Gibbs Sampler for θ

				95% Credible Region	
M	Mean	STD	SEM	Lower	Upper
250	108.4	1.04	0.07	106.03	110.65
500	108.3	0.94	0.04	106.35	110.21
750	108.3	0.9	0.03	106.64	110.15
1500	108.4	0.94	0.02	106.51	110.26

Box & Tiao – The Weighted Mean Problem

target density. They estimated the value of θ to be 108.43. This is close to the value generated using the Gibbs sampler method. The exact posterior distribution of θ is the poly-t-distribution. The effect of m appears to be minimal indicating that 500–750 iterations of the Gibbs sequence are sufficient.

2.9 Comments and Conclusions

Beginning with Bayes theorem, introductory material for the understanding of Bayesian inference is presented in this chapter. Many examples from time series illustrate the Bayesian methodology. For example, Bayesian methods for the analysis of autoregressive time series are analyzed using the theory and methods unique with the Bayesian approach. Inference for the standard populations is introduced. The most useful population for Markov chains is the binomial population, which models the number of one-step transitions among the states of a finite chain. It is shown how to estimate the one-step transition probabilities of the Markov chain, where the posterior distribution of the transition probabilities is a beta. For the analysis of normal time series, the focus is on estimating the variance parameter of the Brownian motion process, where Bayesian estimation and hypothesis testing methods are demonstrated with the gamma distribution. The R package and WinBUGS software allows one to implement the Bayesian methods that are developed in this chapter.

There are many books that introduce Bayesian inference and the computational techniques that will execute a Bayesian analysis, and the reader is encouraged to read Ntzoufras.[27] The material found in Ntzoufras is an excellent introduction to Bayesian inference and to WinBUGS, the computing software that is employed in this book. Also, there are many books on time series. For example, in addition to the standard of Box and Tiao,[10] see Cryer and Chan[28] and Cowpertwait and Metcalfe[29] for two recent books that employ R.

2.10 Exercises

1. For the beta density (2.8) with parameters α and β, show that the mean is $[\alpha/(\alpha+\beta)]$ and the variance is $[\alpha\beta/(\alpha+\beta)^2(\alpha+\beta+1)]$.

2. From Equation (2.14), show the following. If the normal distribution is parameterized with μ and the precision $\tau = 1/\sigma^2$, the conjugate distribution is as follows: (a) the conditional distribution of μ given τ is normal with mean $\bar{x}$ and precision $n\tau$, and (b) the marginal distribution of τ is gamma with parameters $(n-1)/2$ and
$$\sum_{i=1}^{i=n}(x_i-\bar{x})^2/2 = (n-1)s^2/2,$$ where s^2 is the sample variance.

3. Verify Table 2.1, which reports the Bayesian analysis for the parameters of an AR(1) time series.

4. Verify the following statement: To generate values from the $t(n-1,\bar{x},n/S^2)$ distribution, generate values from Student's t-distribution with n 1 degrees of freedom and multiply each by $S/\sqrt{n}$ and then add $\bar{x}$ to each.

5. Verify Equation (2.79), the predictive density of a future observations Z from a normal population with both parameters unknown.

6. Suppose $x_1, x_2, .., x_n$ are independent and that $x_i \sim gamma(\alpha_i, \beta)$, and show that $y_i = x_i/(x_1 + x_2 + \ldots + x_n)$ jointly have a Dirichlet distribution with parameter $(\alpha_1, \alpha_2, \ldots, \alpha_n)$. Describe how this can be used to generate samples from the Dirichlet distribution.

7. Suppose $(X_1, X_2, \ldots, X_k)$ is multinomial with parameters n and $(\theta_1, \theta_2, \ldots, \theta_k)$, where $\sum_{i=1}^{i=k} X_i = n$, $0 < \theta_i < 1$, and $\sum_{i=1}^{i=k} \theta_i = 1$. Show that $E(X_i) = n\theta_i$, $Var(X_i) = n\theta_i(1-\theta_i)$, and $Cov(X_i, X_j) = -n\theta_i\theta_j$. What is the marginal distribution of θ_i?

8. Suppose $(\theta_1, \theta_2, \ldots, \theta_k)$ is Dirichlet with parameters $(\alpha_1, \alpha_2, \ldots, \alpha_k)$, where $\alpha_i > 0$, $\theta_i > 0$, and $\sum_{i=1}^{i=k} \theta_i = 1$. Find the mean and variance of θ_i and covariance between θ_i and θ_j, $i \neq j$.

9. Show the Dirichlet family is conjugate to the multinomial family.

10. Suppose $(\theta_1, \theta_2, \ldots, \theta_k)$ is Dirichlet with parameters $(\alpha_1, \alpha_2, \ldots, \alpha_k)$. Show the marginal distribution of θ_i is beta and give the parameters of the beta. What is the conditional distribution of θ_i given θ_j?

11. For the exponential density

$$f(x|\theta) = \theta \exp -\theta x, \ x>0$$

where x is positive and θ is a positive unknown parameter, suppose the prior density of θ is

$$g(\theta) \propto \theta^{\alpha-1} \exp -\beta\theta, \ \theta>0,$$

what is the posterior density of θ? In Markov jump processes, the exponential distribution is the distribution of the inter arrival times between events.

12. Refer to Section 2.7.2. Based on BUGS code BC 2.1, generate 100 observations from an exponential distribution with mean 3.

13. Refer to Section 2.5.2 on testing hypotheses. Let θ be the autoregressive parameter of an AR(1) time series and consider a test of the null hypothesis $H_0 : \theta = .6$ versus the alternative $H_1 : \theta \neq .6$. Refer to equations (2.53) through (2.58). Use the Bayesian approach of Lee[21] and use the data generated for the AR(1) series.

14. Verify the predictive probability mass function (2.61dx) for the binomial.

15. Verify the predictive distribution (2.79) of m future observations form a Brownian motion process.

16. Replicate the plot of Figure 2.1, where 100 values of the series are generated with the R code.

```
t<-1:100
sig<-0.01
x<-rnorm9n=length(t)-1,sd=sqrt(sig))
x<- c(0,cumsum(x))
plot(t,x, type="1",ylim=c(-2,2))
```

17. Based on the multinomial mass function and R code

```
n<-1000
prob<-c(.2,.2,.2,.2,.2)
multinom(1,n,prob)
```

generate 1000 observations from the multinomial with parameter vector (.2,.2,.2,.2,.2). List the number of counts over the five categories. Do your results imply they were generated from a multinomial with the probabilities (.2,.2,.2,.2,.2)? How would you test the hypothesis that they were?

References

1. Bayes, T. (1764). "An essay towards solving a problem in the doctrine of chances", *Philosophical Transactions of the Royal Society A London*, 53, 370.
2. Laplace, P.S. "Memorie des les probabilities", *Memories de l'Academie des sciences de Paris*, 227, 1778.
3. Stigler, M. (1986). *The History of Statistics. The Measurement of Uncertainty before 1900.* The Belknap Press of Harvard University Press, Harvard.
4. Hald, A. (1990). *A History of Mathematical Statistics from 1750–1930.* Wiley Interscience, London.
5. Lhoste, E. (1923). "Le calcul des probabilities appliqué a l'artillerie, lois de probabilite a prior", *Revu D'artillirie*, Mai, 405-409.
6. Jeffreys, H. (1939). *An Introduction to Probability.* Clarendon Press, Oxford.
7. Savage, L.J. (1954). *The Foundation of Statistics.* John Wiley & Sons Inc., New York.
8. Lindley, D.V. (1965). *Introduction to Probability and Statistics from a Bayesian Viewpoint, Volumes I and II.* Cambridge University Press, Cambridge, UK.
9. Broemeling, L.D. and Broemeling, A.L. (2003). "Studies in the history of probability and statistics XLVIII: The Bayesian contributions of Ernest Lhoste", *Biometrika*, 90(3), 728.
10. Box, G.E.P. and Tiao, G.C. (1973). *Bayesian Inference in Statistical Analysis.* Addison Wesley, Reading, MA.
11. Zellner, A. (1971). *An Introduction to Bayesian Inference in Econometrics.* John Wiley & Sons Inc., New York.
12. Broemeling, L.D. (1985). *The Bayesian Analysis of Linear Models.* Marcel-Dekker Inc., New York.
13. Leonard, T. and Hsu, J.S.J. (1999). *Bayesian Methods. An Analysis for Statisticians and Interdisciplinary Researchers.* Cambridge University Press, Cambridge, UK.
14. Gelman, A., Carlin, J.B., Stern, H.S. and Rubin, D.B. (1997). *Bayesian Data Analysis.* Chapman & Hall/CRC, New York.
15. Congdon, P. (2001). *Bayesian Statistical Modeling.* John Wiley & Sons Ltd, London.
16. Congdon, P. (2003). *Applied Bayesian Modeling.* John Wiley & Sons Inc., New York.
17. Congdon, P. (2005). *Bayesian Models for Categorical Data.* John Wiley & Sons Inc., New York.
18. Carlin, B.P. and Louis, T.A. (1996). *Bayes and Empirical Bayes for Data Analysis.* Chapman & Hall, New York.
19. Gilks, W.R., Richardson, S. and Spiegelhalter, D.J. (1996). *Markov Chain Monte Carlo in Practice.* Chapman & Hall/CRC, Boca Raton, FL.
20. Lehmann, E.L. (1959). *Testing Statistical Hypotheses.* John Wiley & Sons Inc., New York.
21. Lee, P.M. (1997). *Bayesian Statistics, An Introduction, Second Edition.* John Wiley & Sons Inc., New York.
22. Wilks, D. and Wilby, R. (1999). "The Weather Generation Game. A Review of Stochastic Weather Models", *Progress in Physical Geography*, 23, 329–357.
23. Davis, R.A., Holan, S.H., Lund, R. and Ravishanker, N. (2016). *Handbook of Discrete-Valued Time Series.* CRC Press, Chapman & Hall, Boca Raton, FL.

24. Kirch, C. and Kamgaing, J.T. (2016). "Detection of change points in discrete-valued time series", in *Chapter 10 in Handbook of Discrete -Valued Time Series*. edited by Davis, Holan, Lund, and Ravishanker, CRC Press, Boca Raton, FL.
25. Casella, G. and George, E.I. (2004). "Explaining the Gibbs sampler", *The American Statistician*, 46, 167–174.
26. Gregurich, M.A. and Broemeling, L.D.A. (1997). "Bayesian analysis for estimating the common mean of independent normal populations using the Gibbs sampler", *Communications in Statistics*, 26(1), 25–31.
27. Ntzoufras, I. (2009). *Bayesian Modeling Using WinBUGS*. John Wiley & Sons Inc., Hoboken, NJ.
28. Cryer, J.D. and Chan, K.S. (2008). *Time Series Analysis: with Applications in R*. Springer, New York.
29. Cowpertwait, P.S.P. and Metcalfe, A.V. (2009). *Introductory Time Series with R*. Springer, New York.

3

Preliminary Considerations for Time Series

3.1 Time Series

This chapter initiates the study of time series by presenting the necessary concepts that are essential for a good understanding of the subject. It begins with several examples of time series followed by techniques that will allow the investigator to understand the general characteristics of a time series. The first technique is to discern the general appearance of the series using plotting procedures, then from the plot form a subjective opinion of the trend of the series. More formal procedures using R will reveal the various components (e.g., trend and seasonality) of the series. Based on the knowledge of the general appearance of the series, it allows one to propose a tentative model and see how well it fits the series. After a series of diagnostic checks, a final model is selected. Several examples of time series are now presented to illustrate the process of selecting a final model.

3.1.1 Airline Passenger Bookings

According to Cowpertwait and Metcalfe,[1] the following R code will reveal the international airline booking, in thousands per month, for the airline Pan Am. This monthly information is also available from the Federal Aviation Administration for the years 1949–1960. See Table 3.1

RC 3.1.
Data(Air Passengers)
AP<-Air Passengers
AP

The following R command summary provides the basic statistical characteristics (minimum, maximum, first and third quartiles, mean, and median) of the Air Passenger bookings.
   >summary(AP)

TABLE 3.1

	Jan	Feb	Mar	Apr	May	Jun	Jul	Aug	Sep	Oct	Nov	Dec
1949	112	118	132	129	121	135	148	148	136	119	104	118
1950	115	126	141	135	125	149	170	170	158	133	114	140
1951	145	150	178	163	172	178	199	199	184	162	146	166
1952	171	180	193	181	183	218	230	242	209	191	172	194
1953	196	196	236	235	229	243	264	272	237	211	180	201
1954	204	188	235	227	234	264	302	293	259	229	203	229
1955	242	233	267	269	270	315	364	347	312	274	237	278
1956	284	277	317	313	318	374	413	405	355	306	271	306
1957	315	301	356	348	355	422	465	467	404	347	305	336
1958	340	318	362	348	363	435	491	505	404	359	310	337
1959	360	342	406	396	420	472	548	559	463	407	362	405
1960	417	391	419	461	472	535	622	606	508	461	390	432

Min.	1st Qu.	Median	Mean	3rd Qu.	Max.
104.0	180.0	265.5	280.3	360.5	622.0

All information in R is stored in objects, and in this example, the object is class ts (time series), given by the following command:

```
class(AP)
[1] "ts"/.
```

It will be interesting to begin the analysis of this data set by determining its trend and by plotting the data.

3.1.2 Sunspot Data

The book by Feigelson and Babu[2] is an excellent book on the use of statistics in astronomy. For the second example, one of the most interesting time series is the sunspot data.

The 1749 sunspot numbers are shown below, but for the total dataset, the link will provide the sunspot data and the corresponding plot of this interesting series.

http://solarscience.msfc.nasa.gov/SunspotCycle.shtml.

The numbers tabulated in SN_m_tot_V2.0.txt are the monthly averages of the daily sunspot number with error estimates as posted at the WDC-SILSO, Royal Observatory of Belgium, Brussels.

1749 01 1749.042 96.7 −1.0 −1
1749 02 1749.123 104.3 −1.0 −1
1749 03 1749.204 116.7 −1.0 −1
1749 04 1749.288 92.8 -1.0 -1
1749 05 1749.371 141.7 −1.0 −1
1749 06 1749.455 139.2 −1.0 −1
1749 07 1749.538 158.0 −1.0 −1
1749 08 1749.623 110.5 −1.0 −1
1749 09 1749.707 126.5 −1.0 −1
. . ..

Ancient Chinese astronomers as well as Galileo discovered that the sun has small dark spots that emerge on one side and disappear on the other side of the sun as it rotates. The period of this time series is approximately 30 days. According to the famous astronomer George Hale, the sunspots depend on the intensity of magnetic fields appearing on the surface of the sun. Now there is a branch of astronomy that is devoted to the solar physics that studies the morphology, evolution, and implications of these active solar areas. What is observable as a consequence of sunspot are solar flares which produce ener-getic particles and ionic phenomena that are not seen over most of the surface of the sun. Such flares can be disturbing in that they can have a major impact on the weather and on early communication systems. The number of spots observed on the sun and averaged over a month reveals a time series that has a large amplitude with obvious seasonality with a period of approximately 11 years. Refer to Feigelson and Babu[2, pp. 422–423] for a more detailed analysis of sunspot data. The fundamental characteristics (amplitudes, period, and harmo-nic components) of the sunspot cycle will be obvious from plotting the data.

3.1.3 Los Angeles Annual Rainfall

The following time series for the rainfall information can be found by using the link www.stat.uiowa.edu/~kchan/TSA.htm.

TSA is the R package that goes with the book by Cryer and Chan.[3] Go to the website of this book at www.r-project.org.

Time Series:
Start = 1878
End = 1992
Frequency = 1
20.86 17.41 18.65 5.53 10.74 14.14 40.29 10.53 16.72 16.02 20.82 33.26 12.69
12.84 18.72 21.96 7.51 12.55 11.80 14.28 4.83 8.69 11.30 11.96 13.12 14.77
11.88 19.19 21.46 15.30 13.74 23.92 4.89 17.85 9.78 17.17] 23.21 16.67 23.29
8.45 17.49 8.82 11.18 19.85 15.27 6.25 8.11 8.94 18.56 18.63 8.69 8.32 13.02
18.93 10.72 18.76 14.67 14.49 18.24 17.97 27.16 12.06 20.26 31.28 7.40 22.57
17.45 12.78 16.22 4.13 7.59 10.63 7.38 14.33 24.95 4.08 13.69 11.89 13.62 13.24

17.49 6.23 9.57 5.83 15.37 12.31 7.98 26.81 12.91 23.66 7.58 26.32 16.54 9.26
6.54 17.45] 16.69 10.70 11.01 14.97 30.57 17.00 26.33 10.92 14.41 34.04 8.90
8.92 18.00 9.11 11.57 4.56 6.49 15.07 22.65

Thus, the 1878 annual rainfall was 20.86 inches and the 1992 was 22.65.
The plot of this data will reveal the obvious characteristics of the rainfall
including the years with the smallest and largest amounts of rainfall.

3.2 Graphical Techniques

The following three sections give the plots for the three time series
introduced earlier, namely the air passenger booking, the sunspot cycles,
and the Los Angeles annual rainfall amounts.

3.2.1 Plot of Air Passenger Bookings

Use the command plot(AP) to produce the plot of the air passenger data.
See Figure 3.1 below.

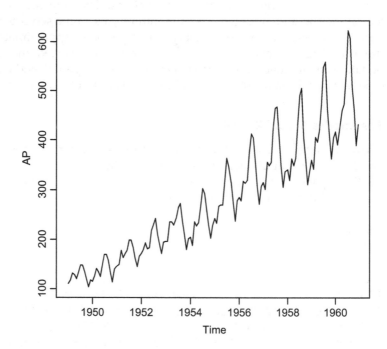

FIGURE 3.1
Airline Passenger Bookings

It is obvious that the trend is increasing, that the series is periodic with an annual period, and the amplitudes are increasing. What will be a suitable model for this time series?

3.2.2 Sunspot Data

See Figure 3.2 below for the sunspot data.

The monthly averages depicted in Figure 3.2 are (updated monthly) the sunspot numbers (181 kb JPEG image), (307 kb pdf-file), (62 kb text file)

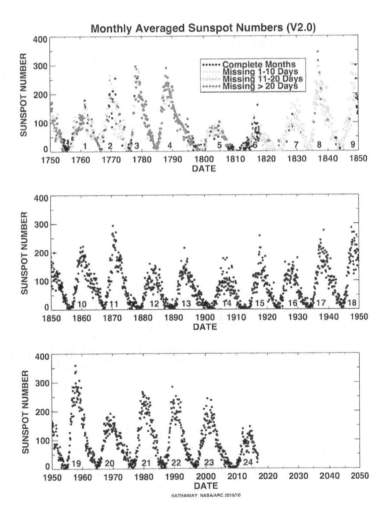

FIGURE 3.2
Monthly Average Sunspot Numbers

that show the number of sunspots visible on the sun waxes and wanes with an approximate 11-year cycle.

(Note: there are actually at least two 'official' sunspot numbers reported. The International Sunspot Number as compiled by the Solar Influences Data Analysis Center in Belgium has been revised recently (V2.0 – summer 2015), and should now more closely match the NOAA sunspot number. The NOAA sunspot number is compiled by the U.S. National Oceanic and Atmospheric Administration. The numbers tabulated in SN_m_tot_V2.0.txt are the monthly averages of the daily sunspot number with error estimates as posted at the WDC-SILSO, Royal Observatory of Belgium, Brussels.)

3.2.3 Graph of Los Angeles Rainfall Data

See Figure 3.3 for the rainfall information.

Figure 3.2 covers more than 100 years and show considerable variation from some very low to other years with quite high readings. For example, a very wet year with 40 inches was recorded for 1883 compared to 1983 with a very dry

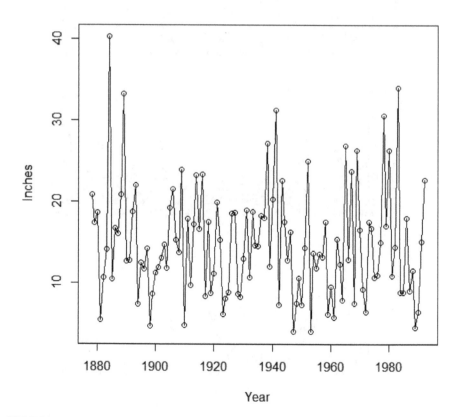

FIGURE 3.3
Los Angeles Rainfall Data

year. Using this information, could one predict future rainfall amounts? This series appears at first glance to be approximately stationary in the mean and variance. The R commands for the above plot use the following R code.

RC 3.2
library (TSA)
data(larain)
plot(larain)
Remember that the package TSA should be downloaded first, then extract the larain dataset.
There does not appear to be any sinusoidal components in this series

---

3.3 Trends, Seasonality, and Trajectories

Based on the plots of a given series, subjective assessments of the components of the series can be proposed. For example, with the Air Passenger bookings, it is apparent that the trend is increasing, and that there is an annual seasonal variation, while for the sunspot time series, the trajectory and overall trend is flat with an obvious seasonal component. Finally, for the Los Angeles rainfall data, the trend is constant, with a large variation in the measured amounts, but there are no seasonal effects.

3.4 Decomposition

Up to this point, the subjective determination of the components of a series has been based on a plot of the data over time. Many time series are dominated by trend with or without seasonal effects.

In this section, an R command is employed to delineate the trend and seasonal effects (if any). This R procedure is based on the additive decomposition model

$$X(t) = m(t) + s(t) + e(t) \tag{3.1}$$

where $X(t)$ is the observed value of the series at time t, $m(t)$ the trend at time t, $s(t)$ the seasonal effect at time t, and $e(t)$ a sequence of correlated random variables with mean zero. However, if the seasonal effects increase with increasing trend, a multiplicative model may be more appropriate, namely

$$X(t) = m(t)s(t) + e(t) \tag{3.2}$$

and if in addition the observed series is positive, the log model

$$\log[X(t)] = m(t) + s(t) + e(t) \tag{3.3}$$

might suffice as an appropriate model.

3.4.1 Decompose Air Passenger Bookings

From Figure 3.1, for the air passenger data, it appears that the trend is increasing so are the seasonal effects; now since the series is positive, the model (3.3) is used for the decompose function in R. See R code 3.3 as follows:

RC 3.3
```
> y<-log(AP)
> plot(y)
>plot(decompose(y))
```

The first command takes the log of the air passengers booking, while the second plots the log values over time, and finally the last plots the decompose of the log series into its various components. For the latter plot, see Figure 3.4(a).

(a)   **Decomposition of additive time series**

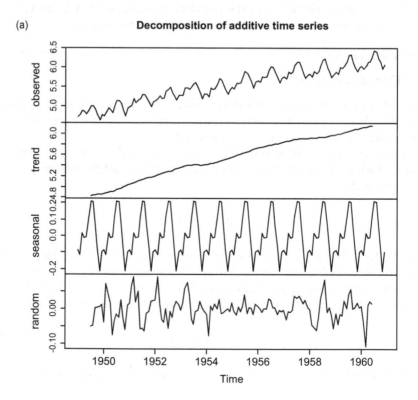

(b)   **Decomposition of additive time series**

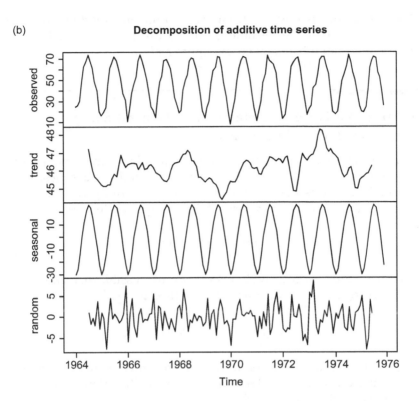

FIGURE 3.4
Decomposition of (a) Air Passenger Bookings; (b) Dubuque Monthly Temperatures

This is an interesting plot because it can be seen that the trend is positive and almost linear, that the annual seasonal effects are systematically repeated and that the random error values have means that appear to be zero; however, I am not sure of the variance of the random errors. Should another more appropriate model be proposed?

The Los Angeles rainfall series does not have obvious cycles and the decompose action for this series will not activate, but will result in the statement 'has no or less than two periods.'

3.4.2 Average Monthly Temperatures of Dubuque Iowa

Another example with apparent periodicity is provided by Cryer and Chan,[3, p. 6] namely the average monthly temperature in degrees Fahrenheit from 1964 to 1976.

Use the command data(tempdub) attached to the R package TSA

The seasonality is obvious from the R command plot(tempdub) and it can be verified that the decompose procedure given by the R code plot (decompose(tempdub)) is given by Figure 3.4(b).

3.5 Mean, Variance, Correlation, and General Sample Characteristic of a Time Series

The mean and variance play an important role in the analysis of time series. The mean of a time series indicates the trend of the series, while the autocorrelation measures the association of neighboring values.

The mean value of a times series $\{X(t), t > 0\}$ is defined as

$$m(t) = E[X(t)], t > 0 \tag{3.4}$$

where the expectation is with respect to all possible realizations of the process. Also, the variance function is defined as

$$Var(t) = E[X(t) - m(t)]^2, t > 0. \tag{3.5}$$

Thus, for the variance function to exist the mean value function must exist. Again, the expectation involved in the variance function is taken over all possible realizations of the series.

Consider a time series that is stationary in the mean and variance where the values may be correlated, then the second-order properties of the series can be measured by the autocovariance

$$\gamma(k) = E[X(t+k) - \mu)][X(t) - \mu], t > 0, k = 0, 1, 2, \ldots \tag{3.6}$$

where k is called the lag and the mean is

$$\mu = [X(t)], t > 0. \tag{3.7}$$

Note the mean value function (3.4) has the constant value μ.
Now the autocorrelation function can be defined as

$$\rho(k) = \gamma(k)/\sigma^2, \tag{3.8}$$

where σ^2 is the constant variance

$$\sigma^2 = \gamma(0). \tag{3.9}$$

Note the variance (3.9) is constant because it is assumed the series is second-order stationary!

The usual estimator of the autocovariance (3.6) is lag k

$$c(k) = (1/n) \sum_{t=1}^{t=n-k} [x(t+k) - \bar{x}][x(t) - \bar{x}] \qquad (3.10)$$

called the sample covariance at lag k where the sample information is denoted by $\{x(t), t = 1, ..., n-k\}$ and $\bar{x} = \sum_{t=1}^{t=n} x(t)$ is the sample mean. Note that $c(0)$ is the sample variance. This is sufficient information to define the sample autocorrelation as

$$r(k) = c(k)/c(0). \qquad (3.11)$$

The Bayesian approach to estimating the first- and second-order properties will be based on different considerations.

The autocorrelation function and its graph (the correlogram) will be demonstrated with the Los Angeles rainfall data recorded annually for 100 years beginning in 1883. This information is available at Cryer and Chan[3, p. 2]

20.86, 17.41, 18.65, 5.53, 10.74, 14.14, 40.29, 10.53, 16.72, 16.02, 20.82, 33.26, 12.69, 12.84, 18.72, 21.96, 7.51, 12.55, 11.80, 14.28, 4.83 8.69, 11.30, 11.96, 13.12, 14.77, 11.88, 19.19, 21.46, 15.30, 13.74, 23.92, 4.89, 17.85, 9.78, 17.17, 23.21, 16.67, 23.29, 8.45, 17.49, 8.82, 11.18, 19.85, 15.27, 6.25, 8.11, 8.94, 18.56, 18.63, 8.69, 8.32, 13.02, 18.93, 10.72, 18.76, 14.67, 14.49, 18.24, 17.97, 27.16, 12.06, 20.26,31.28, 7.40, 22.57, 17.45, 12.78, 16.22, 4.13, 7.59,10.63, 7.38, 14.33, 24.95, 4.08, 13.69, 11.89, 13.62, 13.24, 17.49, 6.23, 9.57, 5.83,15.37, 12.31, 7.98, 26.81, 12.91, 23.66, 7.58,26.32, 16.54, 9.26, 6.54, 17.45, 16.69, 10.70, 11.01, 14.97, 30.57, 17.00, 26.33, 10.92, 14.41, 34.04, 8.90, 8.92, 18.00, 9.11, 11.57, 4.56, 6.49,15.07, 22.65,

The r command for the plot of the rainfall data is
>plot(larain)
and the graph appears in Figure 3.5. It appears that the series is approximately stationary with mean estimated as mean(larain) = 14.88843 inches per year, and standard deviation estimated by sd(larain)
6.880769 inches.

The correlogram for this series appears in Figure 3.6. In particular, the lag five correlation is estimated with the r command acf(larain)$ac[5]= 0.1720229, which is also demonstrated by Figure 3.6.

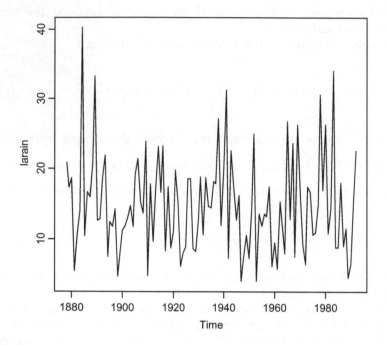

FIGURE 3.5
Los Angeles Rainfall Data

3.6 Other Fundamental Considerations

One of the fundamental concepts in the analysis of time series is model building. This comprises model specification, model fitting, and model diagnostics. After discerning the general characteristic of a series and choosing a tentative model, one must assess how well the model fits the data. How well the model fits the data is usually based on the residuals, and for the Bayesian depends on the Bayesian predictive density. See Cryer and Chan for additional details about diagnostic checks for fitting time series.

3.7 Summary and Conclusions

This chapter briefly describes the fundamental concepts for selecting a model that gives a good representation of the observations. First, one plots to reveal the general overall appearance of the series, but it is the decompose R command that provides the researcher with a definite revelation of the various components (trend, seasonality, and error

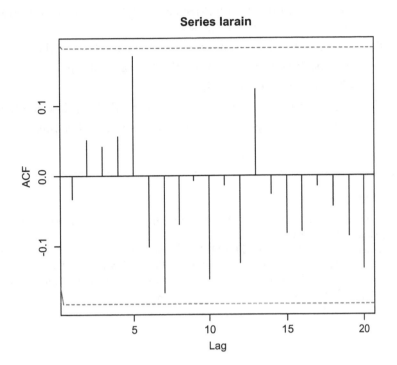

FIGURE 3.6
Correlogram of Los Angeles Rainfall Data

structure) of the series. The decompose action allows one to propose a tentative model, followed by a series of diagnostic checks involving the residuals that allow one to select a final model. Many examples provide an illustration of the steps involved in the analysis of time series.

3.8 Exercises

1. (a) Refer to Section 3.1.1 and the airline passenger booking data, then using R, plot the data.
   (b) Using the plot(decompose (y)) R command, implement the decompose procedure.
   (c) What does the decompose command reveal about the overall appearance of the airline data?
2. Refer to Section 3.1.2, the sunspot cycle data.
   (a) Use the R command plot to plot the sunspot cycle data.

(b) After executing the plot(decompose(y)) command, describe the overall appearance of the series.

3. Refer to Section 3.1.3 and annual Los Angeles Rainfall data.

   (a) Use the R command plot to plot the rainfall data.

   (b) What is the trend for this series?

   (c) Are there seasonal effects for this series? If so, describe them.

4. Write a brief two-page essay about the content of Chapter 3.

References

1. Cowpertwait, P.S. and Metcalfe, A.V. (2009). *Introductory Time Series with R*. Springer, New York.
2. Feigelslon, E.D. and Babu, G.J. (2012). *Modern Statistical Methods for Astronomy, with R Applications*. Cambridge University Press, Cambridge, UK.
3. Cryer, J.D. and Chan, K.S. (2008). *Time Series Analysis: with Applications in R*. Springer, New York.

4

Basic Random Models

4.1 Introduction

Up to this point, two approaches for modeling time series have been discussed. As previously explained, the first is founded on the belief that there is a fixed seasonal pattern about a trend, such as the Air Passenger data and these two features can be delineated with the R decompose command. The second approach takes account of the seasonal pattern and trend to change over time. With mathematical time series, the difference between the fitted value and the observed value allows one to calculate the random error series. If the model includes various aspects of the deterministic characteristic of the series, then the residuals should exhibit as a realization of independent random variables, but this may not be the case. That is to say, the residuals may show some structure such as a trend or positive autocorrelation.

4.2 White Noise

Because a good fit implies that the residuals are independent random variables, the model to be presented next will be with a foundation of white noise.

White noise is a time series

$$\{W(t), t = 1, 2, ..., n\} \tag{4.1}$$

of variables $W(1), W(2), ..., W(n)$, which are independent and identically distributed with mean 0, constant variance σ^2, and of course $cor[W(i), W(j)] = 0, i \neq j$. In addition, if $W(i) \sim N(0, \sigma^2)$, the noise is referred to Gaussian or normal white noise.

R is useful to simulate time series and this will be done for the basic stochastic models, such a white noise, random walks, and random walks with drift. Consider the scenario where a fitted time series can be used to simulate data. As has been seen throughout this book, simulation is used for a variety of reasons. In R, simulation is a simple operation where most of the well-known distributions are simulated with a R function. For example, for a simulation of normal random variables, rnorm(100) generates 100 standard normal variables. Now consider, the code

RC 4.1.
set.seed(1)
w<-rnorm(100)
time<-seq(1,100,1)
plot(time,w)

that generates 100 white noise values and the plot abscissa has a unit of one. One should check to see how well the random number generator simulates white noise. For example, the sample mean(w)=.10887, sample standard deviation sd(w)=.8067, and the lag one correlation is acf(w)$acf[2] =−.00365, and finally, the lag two is given by acf(w)$acf[3]=−.02707. Of course, one should also use the command acf(w) to plot the autocorrelation function of the w series. The command hist(w) generates the default histogram of the w purported white noise series and is valuable in detecting departures from a normal distribution. Such evaluations are necessarily subjective, in that another individual might detect different deviations from normality.

The next stochastic model to be studied is the random walk series

---

4.3 A Random Walk

The next stochastic model to be studied is the random walk series

$$X(t) = X(t-1) + W(t), t = 1, 2, \ldots \tag{4.2}$$

where $\{W(t), t = 1, 2,\}$ is a white noise process. Another representation of the random walk is an infinite series

$$X(t) = W(t) + W(t-1) + W(t-2) + \ldots,$$

and since the series begins at some point say $t= 1$, it can be expressed as

$$X(t) = W(1) + W(2) + \ldots + W(t). \tag{4.3}$$

One may show that the first and second moments are

$$\mu(t) = E[X(t)] = 0,$$

$$\gamma_k(t) = Cov[X(t), X(t+k)] = t\sigma^2 \qquad (4.4)$$

and is a function of t, and thus the process is not covariance stationary. Also, the autocorrelation is

$$\rho_k(t) = Cov[X(t), X(t+k)]/(\sqrt{t\sigma^2(t+k)\sigma^2}$$

$$= 1/\sqrt{1+k/t} \qquad (4.5)$$

thus, the autocorrelation function is positive and decays from 1 to 0. Consider a simulation of a random walk with R with code

RC 4.2
```
x<-w<-rnorm(1000)
for ( t in 2·1000) x[t]<-x[t-1]+w[t]
time<-1:1000
plot(time,x)
```
and the plot of the random walk values over 1000 time points portrayed in Figure 4.1. I computed the following autocorrelations: the lag two as acf(x)$acf[3]=.9901 and lag 28 as acf$acf[29]=.87105 which follows from formula (4.2). The student will be asked to display the autocorrelation function with the code acf(x). The random walk values start at zero, increase to 9, then rapidly fall to an average value of −25. The average value of the 1000 values is −10.92, which is reasonable after looking at Figure 4.2.

Consider a more general random walk model with drift, namely

$$X(t) = X(t-1) + \zeta + W(t), t = 1, 2, ..., n \qquad (4.6)$$

where $W(t)$ is Gaussian white noise with variance σ^2.

One hundred values for this random walk with drift $\zeta = 3$ and variance $\sigma^2 = 1$ are simulated with the following R code.

RC 4.3
```
delta=3
x<-w<-d<-rnorm(100)
for (t in 2:100) {x[t]<-x[t-1]+ delta+w[t]
d[t]<-x[t]-x[t-1]}
time<-1:100
plot(time,x)
```
The 100 increments d(t) are the components of the vector d:

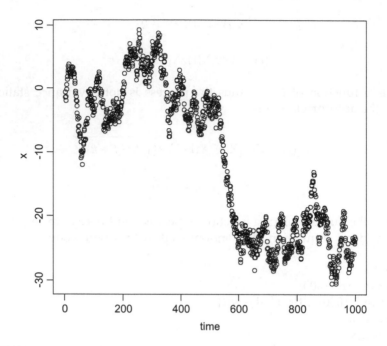

FIGURE 4.1
Graph of a Random Walk

d=c(0.7140855, 3.5813846, 2.8532761, 4.5069818, 2.7204674, 5.0277387,
1.8042598, 4.3123179, 2.4759925, 3.3542495, 2.9283255, 2.8668557, 2.9119226,
3.9177889, 3.0314393, 4.3589306, 3.1134948, 3.1743065, 2.9465196, 2.5752420,
3.2079116, 3.9929229, 3.8138579, 3.9960056, 4.7327443, 3.3570420, 3.1557263,
4.5527887, 2.9359304, 1.9917812, 2.1071045, 2.3454854, 1.5662172, 4.4181844,
2.4061143, 4.1189718, 5.3840816, 2.5967135, 3.3701086, 2.4247070, 5.9192288,
1.9210852,] 4.1784913, 3.8696028, 3.9311663, 4.0246812, 3.6129751, 3.6081901,
2.1071083, 3.4520356, 2.6252877, 5.8401851, 3.7323935, 3.8405189, 0.7803479,
1.6725362, 2.0233393, 2.4129241, 2.4319872, 3.6436410, 2.7626929, 2.7014220,
4.1895626, 0.8910588, 2.9267180, 3.8634029, 4.0747075, 2.7941474, 2.5663862,
2.8177762, 2.5689213, 2.5753904, 1.6461803, 2.8034051, 3.2341161, 2.4427937,
1.8866747, 1.3693646, 2.6121761, 4.7226320, 2.4414997, 2.3633392, 5.7813871,
1.0427441, 3.1425279, 3.6644988, 2.5213725, 4.1836896, 3.8054373, 3.0533178,
1.8326536, 3.3233502, 2.6792170, 2.1140497, 3.1512782, 3.5793635, 2.9208332,
3.8487676, 0.4889665, 3.7384771).

The goal is to estimate the drift and variance based on the following
differences:

$$d(t) = X(t) - X(t-1), t = 2, 3, \ldots, n \qquad (4.7)$$

where $d(t) = \zeta + W(t)$; therefore, $E[d(t)] = 3$, $Var[d(t)] = \sigma^2$, and $Cov[d(s), d(t)] = 0$.

For the Bayesian analysis, assign noninformative priors for the mean μ and variance σ^2, namely for μ a normal $(0,.001)$ distribution and for $\tau = 1/\sigma^2$ a gamma$(.001,.001)$ distribution. The Bayesian analysis is executed with 35,000 for the simulation and 5,000 for the burn-in.

BC 4.1.
```
model;
{
# the prior distributions
mu~dnorm(0,.001)
tau~dgamma(.001,.001)
for (i in 1:100){
d[i]~dnorm(mu, tau)
z[i]~dnorm(mu,tau)}
sigma<-1/tau
}
list( d=c( 0.7140855, 3.5813846, 2.8532761, 4.5069818, 2.7204674, 5.0277387,
1.8042598, 4.3123179, 2.4759925, 3.3542495, 2.9283255, 2.8668557, 2.9119226,
3.9177889, 3.0314393, 4.3589306, 3.1134948, 3.1743065, 2.9465196, 2.5752420,
3.2079116, 3.9929229, 3.8138579, 3.9960056, 4.7327443, 3.3570420, 3.1557263,
4.5527887, 2.9359304, 1.9917812, 2.1071045, 2.3454854, 1.5662172, 4.4181844,
2.4061143, 4.1189718, 5.3840816, 2.5967135, 3.3701086, 2.4247070, 5.9192288,
1.9210852, 4.1784913, 3.8696028, 3.9311663, 4.0246812, 3.6129751, 3.6081901,
2.1071083, 3.4520356, 2.6252877, 5.8401851, 3.7323935, 3.8405189, 0.7803479,
1.6725362, 2.0233393, 2.4129241, 2.4319872, 3.6436410, 2.7626929, 2.7014220,
4.1895626, 0.8910588, 2.9267180, 3.8634029, 4.0747075, 2.7941474, 2.5663862,
2.8177762, 2.5689213, 2.5753904, 1.6461803, 2.8034051, 3.2341161, 2.4427937,
1.8866747, 1.3693646, 2.6121761, 4.7226320, 2.4414997, 2.3633392, 5.7813871,
1.0427441, 3.1425279, 3.6644988, 2.5213725, 4.1836896, 3.8054373, 3.0533178,
1.8326536, 3.3233502, 2.6792170, 2.1140497, 3.1512782, 3.5793635, 2.9208332,
3.8487676, 0.4889665, 3.7384771))
list(mu=0,tau=1)
```

The posterior analysis is reported in Table 4.1.

Recall that the value of μ used to simulate the differences $d(t)$ is 3 (the drift) and that for τ is 1. Thus, according to Table 4.1, it seems that the simulated values are believable. For example, the posterior mean of μ is 3.104 with a 95% credible interval of (2.887,3.321) which includes 3. In a similar way, the posterior mean of σ^2 is 1.204 with a 95% credible interval (.91,1.597) which included the value 1!

TABLE 4.1

Posterior Analysis for Random Walk with Drift

Parameter	Mean	SD	Error	2½	Median	97½
μ	3.104	.1103	.000644	2.887	3.104	3.321
σ^2	1.204	.1748	.000948	.91	1.189	1.597
τ	.8474	.1203	.000652	.6263	.841	1.099

See Cowpertwait and Metcalfe[1,chap 4] for more information about using R to generate observations from these basic time series models.

4.4 Autoregressive Models

The next class of basic time series models is the autoregressive process $AR(p)$ defined as

$$Y(t) = \sum_{i=1}^{i=p} \theta_i Y(t-i) + W(t) \tag{4.9}$$

Or in terms of the backshift operators

$$\Phi_p(B)Y(t) = (1 - \theta_1 B - \theta_2 B^2 - ...\theta_p B^p)Y(t) + W(t),$$

where B is the backshift operator defined as

$$B^m Y(t) = Y(t-m).$$

Also, $\{W(t), t>0\}$ is Gaussian white noise with variance σ^2, and the $\theta_i, i = 1, 2, ..., p$ a sequence of unknown autoregressive parameters.

It should be observed that the random walk is a special case of the AR(1) process with $\theta_1 = 1$, and the class is called autoregressive because $Y(t)$ is regressed on past terms of the same process.

AR processes can be stationary or not, depending on the roots of the characteristic equation $\Phi_p(B) = 0$. Treat the operator B as a real or complex number, then if the roots of the characteristic equation all lie outside the unit circle, then the process is stationary.

Note the characteristic equation

$$\Phi_p(B) = 0 \tag{4.10}$$

Is

where the p-th order polynomial in B is

$$\Phi_p(B) = (1 - \theta_1 B - \theta_2 B^2 - \ldots - \theta_p B^p).$$

The following three examples show how to determine if an AR process is stationary.

(a) The AR(1) process Y(t)=(1/2)Y(t-1)+W(t) is stationary because the root of 1-B/2=0 is B=2, which is greater than 1! An more generally, the AR(1) process is stationary if $|\theta_1| < 1$.

(b) Also, the AR(2) process Y(t)=Y(t-1)-(1/4)Y(t-2)+W(t) is stationary because the roots of the characteristic equation

$(1/4)(B-2)^2 = 0$ has two roots which are both B=2.

(c) On the other hand, the AR(2) process
Y(t)= (1/2)Y(t-1)+(1/2)Y(t-2)+W(t) is not stationary. This is left as an exercise for the reader.

For the AR(1) model, $Y(t) = \theta Y(t-1) + W(t)$, the first two moments are

$$\mu(t) = 0$$

and covariance function with lag $k = 1,2,\ldots$

$$\gamma_k = \theta^k \sigma^2 / (1 - \theta^2), \tag{4.11}$$

and also with autocorrelation function

$$\rho_k = \theta^k, \ |\theta| < 1 \ldots \tag{4.12}$$

It is obvious from (4.12) that the autocorrelations are non-zero and decay exponentially with k. Another important second-order property of the AR(p) process is the partial correlation function at lag k is the correlation that results after removing the effects of correlations of terms with lags less than k.

The following R code RC 4.4 is given by Cowpertwait and Metcalfe[1,p. 81] which executes 100 values from the AR(1) process with autoregression coefficient $\theta = .6$. The autocorrelation function is acf while that for the partial autocorrelation is pacf.

RC 4.4
```
set.seed(1)
y<-w<-rnorm(100)
for( t in 2:100) y[t]<-.6*y[t-1]+w[t]
```

time <- 1:100
plot(time,y)
acf(y)
pacf(y)
The first 50 values are labeled by the vector y= (−0.62645381, −0.19222896, −0.95096599, 1.02470121, 0.94432850, −0.25387129, 0.33510628, 0.93938847, 1.13941444, 0.37826027, 1.73873733 1.43308564, 0.23861080, −2.07153341, −0.11798913, −0.11572708, −0.08562651, 0.89246030, 1.35669738, 1.40791975, 1.76372922, 1.84037383, 1.17878928, −1.28207813, −0.14942113, −0.14578142, −0.24326436, −1.61671100, −1.44817665, −0.45096443, 1.08810089, 0.55007281, 0.71771530, 0.37682414, −1.15096507, −1.10557361, −1.05763412, −0.69389387, 0.68368905, 1.17338918, 0.53950991, 0.07034427, 0.73916994, 1.00016516, −0.08865660, −0.76068912, −0.09183151, 0.71343402, 0.31571420, 1.07053625)

The posterior density of θ is derived as follows.
The likelihood function is

$$l(\theta|y) = (\tau/2\pi)^{n/2} \exp -(\tau/2) \sum_{t=2}^{t=n} [y(t) - \theta y(t-1)]^2, \qquad (4.13)$$

where y is the vector of observations $y = (y(1), y(2), \ldots, y(n))$ and $|\theta| < 1$. Suppose the prior density of θ is uniform $\zeta(\theta) = 1$, and for τ

$$\varsigma(\tau) = 1/\tau,$$

$$\tau > 0.$$

It can be shown that the posterior density of θ is

$$f(\theta|data) \propto f(\theta|data) \propto \left\{1 + \lambda[\theta - \nu]^2\right\}^{-n/2} \qquad (4.14)$$

where $\nu = [\sum_{t=2}^{t=n} y(t)y(t-1)]/\sum_{t=2}^{t=n} y^2(t)$. This is the density of a t-distribution with $n-1$ degrees of freedom, mean ν, and precision $\lambda = (n-1)/c$, where $c = \sum_{t=1}^{t=n} y^2(t) - [\sum_{t=2}^{t=n} y(t)y(t-1)]^2 / \sum_{t=2}^{t=n} y^2(t-1)$.

I computed the following: $\sum_{t=1}^{t=n} y^2(t) = 46.617288$, $\sum_{t=2}^{t=n} y^2(t-1) = 46.22491$, and $\sum_{t=2}^{t=n} y(t)y(t-1) = 28.87423$, thus, $c=28.58109$, and $\lambda = 1.71442$. The posterior distribution of θ is a univariate t with 49 degrees of freedom, mean $\nu = .62464$, and precision $\lambda = 49/28.58109 = 1.7144$. The posterior mean of

θ is .62464 and compares favorably with the value $\theta = .6$ used to generate the data vector y generated by **RC 4.4**.

Note that the AR(1) process is represented as

$$Y(t) = \theta Y(t-1) + W(t) \tag{4.15}$$

where $\{W(t), t > 0\}$ is Gaussian white noise with variance σ^2, then its mean is 0 and the autocovariance function is

$$\gamma(k) = \theta^k \sigma^2 / (1 - \theta^2), \tag{4.16}$$

and the variance

$$V[Y(t)] = \sigma^2 / (1 - \theta^2) \tag{4.17}$$

Based on the representation (4.15) of an autoregressive process, the Bayesian analysis is repeated, with **BC 4.2** executed with 35,000 observations for the simulation and a burn-in of 5,000. Note the prior for θ is beta (6,4) and that for the variance is expressed as a gamma(.001,.001) distribution. This induces a prior for variance covariance matrix of the 50 observations Y. The code is based on the fact that the 1 by 50 vector Y has a multivariate normal distribution with mean vector zero and covariance matrix given by (4.15) and (4.16). The Bayesian analysis based on WinBUGS is reported in Table 4.2.

BC 4.2

```
model;
{

v~dgamma(.01,.01)
theta~dbeta(6,4)

Y[1,1:50] ~ dmnorm(mu[], tau[, ])
    for( i in 1:50){mu[i]<-0}


    tau[1:50,1:50]<-inverse(Sigma[,])


    for(i in 1:50){Sigma[i,i]<-v/(1-theta*theta)}
        for(i in 1:50){for(j in i+1:50){Sigma[i,j]<- v*pow(theta,j)*1/(1-theta*theta)}}
        for( i in 2:50){ for(j in 1: i-1){Sigma[i,j]<-v*pow(theta,i-1)*1/(1-theta*theta)}}
```

sf.3<-v/(1+pow(theta,2)−2*theta*cos(2*3.416*0.3))

}

list(Y= structure(.Data=c(−0.62645381, −0.19222896, −0.95096599, 1.02470121,
0.94432850, −0.25387129, 0.33510628, 0.93938847, 1.13941444, 0.37826027,
1.73873733, 1.43308564, 0.23861080, −2.07153341, −0.11798913, −0.11572708,
−0.08562651, 0.89246030, 1.35669738, 1.40791975, 1.76372922, 1.84037383,
1.17878928, −1.28207813, −0.14942113, −0.14578142, −0.24326436, −1.61671100,
−1.44817665, −0.45096443, 1.08810089, 0.55007281, 0.71771530, 0.37682414,
−1.15096507, −1.10557361, −1.05763412, −0.69389387, 0.68368905, 1.17338918,
0.53950991, 0.07034427, 0.73916994, 1.00016516, −0.08865660, −0.76068912,
−0.09183151, 0.71343402, 0.31571420, 1.07053625),.Dim=c(1,50)))
list(theta=.6,v=1)

The θ parameter is the main parameter of interest and is the correlation
between observations spaced one unit apart and is estimated as .6258 with
the posterior mean and 95% credible interval (.3096, .8683) which contains
the value $\theta = .6$ used to generate the data. Also, remember that that $\sigma^2 = 1$
is the value used to generate the data which compares to a posterior mean
.5724 and 95% credible interval (.228, 1.043).

R can perform the analysis of an AR process with maximum likelihood
estimation. Consider the 100 observations of an AR(1) process generated
by RC 9.7, then the following commands invoke the ar function followed
by the code for estimating the order, which is 1.

> y.ar<-ar(y,method="mle")
> y.ar$order
[1] 1
The following command estimates the coefficient θ via maximum likelihood as .52311.
> y.ar$ar
[1] 0.5231187
This command generates a 95% confidence interval for θ
> y.ar$ar+c(−2,2)*sqrt(y.ar$asy.var)
(0.3521863, 0.6940510).

TABLE 4.2

Posterior Analysis for AR(1)

Parameter	Mean	SD	Error	2½	Median	97½
θ	.6258	.1509	.002015	.3096	.6383	.8683
σ^2	.5724	.2138	.002894	.228	.552	1.043

Note the difference in the MLE of .523 compared to the posterior mean of .5993. Why the difference? This is let as an exercise at the end of the chapter.

4.5 Another Example

Consider another AR(1) series (4.14) with $\theta = .9$ and $\sigma^2 = .2$. This example is taken from Cryer and Chan[2,p. 68] and the following R code generates 50 observations from this series.

RC 4.5.

```
data(ar1.s)
> ar1.s
Time Series:
Start = 1
End = 50
Frequency = 1
```
−1.8893704, −1.6906396, −1.9621019, −0.5664021, −0.6272304, −0.5262976, 0.7211378, 0.9929824, 0.5646112, 2.1790080, 1.0430490, 0.8509367, 2.0861368, 3.6083092, 5.4100744, 4.5533345, 3.5229049, 1.7642580, 3.8556920, 2.6992689, 2.8096577, 3.1338287, 3.8401200, 3.9310510, 1.3519995, 2.1499917, 2.4104280, 2.5596668, 1.5763719, 2.4052907, 3.5887452, 3.7146012, 3.6923775, 4.1832640, 4.1695488, 4.1196388, 2.1884759, 3.5194316, 3.6673695, 3.7615054, 5.4620648, 4.6082078, 5.0997579, 5.1225704, 4.5147656, 3.9212257, 2.3460774, 2.6567881, −0.1907462, 0.6072285.

The graph of these fifty values is shown in Figure 4.2.

Using this information as the data for BC 4.2, the following Table 4.3 gives results of executing BC 4.2 with 5000 initial observations for the simulation of 40,000 values.

It is interesting to note that the 95% credible interval for θ does not include .9. Why? The student will be asked to give several reasons why this is true.

4.6 Goodness of Fit

How well the model fits the data depends on the Bayesian predictive distribution. For the autoregressive AR(1) model (4.14), the Bayesian predictive density for m future observations $z = (z_{n+1}, z_{n+2}, ...z_{n+m})$ is

$$f(z|y) = \int_{-\infty}^{\infty} \int_{0}^{\infty} f(z|\theta, \tau)\lambda(\theta, \tau|y)d\theta d\tau; \tag{4.18}$$

where

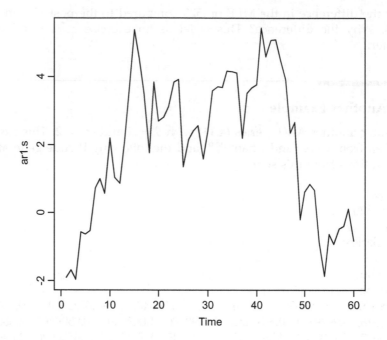

FIGURE 4.2
AR(1) values $\theta = .9$

TABLE 4.3

Posterior Analysis for AR(1)

Parameter	Mean	SD	Error	2½	Median	97½
θ	.9969	.007543	.0000603	.9701	.9878	.9985
σ^2	.2008	.04997	.000424	.126	.1932	.3188

$$f(z|\theta, \tau) = f(z|\theta, \tau) \propto \tau^{m/2} \exp -(\tau/2) \sum_{t=n+1}^{n+m} [z(t) - \theta z(t-1)]^2 \qquad (4.19)$$

And the posterior density of τ and θ is

$$\lambda(\theta, \tau|y) = \lambda(\theta, \tau|y) \propto \tau^{n/2-1} \exp -(\tau/2) \sum_{t=2}^{t=n} [y(t) - \theta y(t-1)]^2 \qquad (4.20)$$

It can be shown that the predictive density is an m-dimensional multivariate t density. This is left as an exercise for the student. See Chapter 5 of Broemeling[3,pp. 186–187] for a derivation of the predictive density.

The following WinBUGS code generates 50 predicted values corresponding to the 50 observations of the AR(1) example of Section 4.5. The predicted values are given by the vector z in the code.

BC 4.3.

```
{

v~dgamma(.01,.01)
theta~dbeta(9,1)

Y[1,1:50] ~ dmnorm(mu[], tau[,])
Z[1,1:50] ~ dmnorm(mu[],tau[,])
X<-Y[1,1:50]-Z[1,1:50]
    for( i in 1:50){mu[i]<-0}


    tau[1:50,1:50]<-inverse(Sigma[,])


    for(i in 1:50){Sigma[i,i]<-v/(1-theta*theta)}
        for(i in 1:50){for(j in i+1:50){Sigma[i,j]<- v*pow(theta,j)*1/
(1-theta*theta)}}
        for( i in 2:50){ for(j in 1: i-1){Sigma[i,j]<-v*pow(theta,i-1)*1/
(1-theta*theta)}}
    }

list(Y= structure(.Data=c( −1.8893704, −1.6906396, −1.9621019, −0.5664021,
−0.6272304, −0.5262976, 0.7211378, 0.9929824, 0.5646112, 2.1790080,
1.0430490, 0.8509367, 2.0861368, 3.6083092, 5.4100744, 4.5533345, 3.5229049,
1.7642580, 3.8556920, 2.6992689, 2.8096577, 3.1338287, 3.8401200, 3.9310510,
1.3519995, 2.1499917, 2.4104280, 2.5596668, 1.5763719, 2.4052907, 3.5887452,
3.7146012, 3.6923775, 4.1832640, 4.1695488, 4.1196388, 2.1884759, 3.5194316,
3.6673695, 3.7615054, 5.4620648, 4.6082078, 5.0997579, 5.1225704, 4.5147656,
3.9212257, 2.3460774, 2.6567881,−0.1907462, 0.6072285 ),.Dim=c(1,50)))
list(theta=.9,v=1)
```

The x vector in the above code is the vector of 50 differences between the observed and predicted observations. The Bayesian analysis for these residuals gives the posterior mean, standard deviation, median, and 95% credible interval. See Table 4.4.

It is seen that the posterior average of the residuals is −1.907 with a 95% credible interval of (−9.017,5.042). One would expect the credible interval to include 0 if the model is a good fit to the AR(1) model. Of course, a good diagnostic analysis should plot the observed values versus the corresponding predicted values.

TABLE 4.4

Posterior Analysis for the Residuals

Parameter	Mean	SD	Error	2½	Median	97½
X	−1.907	3.723	.02215	−9.017	−1.907	5.042

4.7 Predictive Distributions

The Bayesian approach is with the predictive distribution, which will be derived for the AR(p) time series with posterior density i

$$\xi(\theta, \tau|s_n) \propto \tau^{(n+2\alpha+p)/2} \exp(-\tau/2)[(\theta - A^{-1}C)'A(\theta - A^{-1}C) + D], \quad (4.21)$$

where the prior distributions are given by:

$$\zeta(\theta|\tau) \propto \tau^{p/2} \exp -(\tau/2)(\theta - \mu)'P(\theta - \mu) \quad (4.22)$$

and

$$\zeta(\tau) \propto \tau^{\alpha-1}e^{-\tau\beta}. \quad (4.23)$$

Also, s_n is the vector of n observations,

$$A = P + (a_{ij}), \quad (4.24)$$

$$C = P\mu + (c_j), \quad (4.25)$$

and

$$D = 2\beta + \mu'P\mu + \sum_{t=1}^{t=n} Y^2(t) - C'A^{-1}C, \quad (4.26)$$

where the ij component of the A matrix is

$$a_{ij} = \sum_{t=1}^{t=n} Y(t-i)Y(t-j) \quad (4.27)$$

and

$$c_j = \sum_{t=1}^{t=n} Y(t)Y(t-j).$$

In addition, the j-th component of the px1 column C is

$$c_j = \sum_{t=1}^{t=n} Y(t)Y(t-j). \tag{4.28}$$

There is sufficient information to derive the predictive density, which is the conditional density of the future observation $Y(n+1)$ given the past observations s_n. Consider the conditional density $Y(n+1)$ given s_n, θ, and τ

$$f[Y(n+1)|s_n, \theta, \tau] \propto \tau^{1/2} \exp -(\tau/2)[Y(n+1) \\ - \theta_1 Y(n) - \theta_2 Y(n-1) - \dots - \theta_p Y(n-(p-1)), \tag{4.29}$$

which can be expressed (after completing the square in θ and eliminating τ via integration) as

$$f[Y(n+1)|s_n, \theta, \tau] \propto \tau^{1/2} \exp -(\tau/2)[Y^2(n+1) + \theta'E\theta - 2\theta'F], \tag{4.30}$$

where E is the pxp matrix with ij-th element $e_{ij} = Y(n-i+1)Y(n-j+1)$, and F is a px1 vector with j-th component $Y(n+1)Y(n-j+1)$.
Expression (4.30) can be simplified to

$$g[Y(n+1)|s_n) \propto [Y^2(n+1) + D + C'A^{-1}C - (C+F)'(A+E)^{-1} \\ (C+F)]^{-(n+2\alpha+1)/2} \tag{4.31}$$

This in turn can be simplified to show that the predictive distribution of $Y(n+1)$ is a univariate t with $n+2\alpha$ degrees of freedom, location,

$$AVG[Y(n+1)|s_n] = [1 - F^*(A+E)^{-1}F^*]^{-1}F^{*'}(A+E)^{-1}C \tag{4.32}$$

and precision

$$Pr\ e[Y(n+1)|s_n] = [1 - F^*(A+E)^{-1}F^*]\{D + C'A^{-1}C - C'(A+E)^{-1}F^* \\ [1 - F^*(A+E)^{-1}F^*]^{-1}F^{*'}(A+E)^{-1}C\}. \tag{4.33}$$

Note that F^* is a px1 vector with j-th component $Y(n-(j-1))$ and $F=Y(n+1) F^*$.
For our example of forecasting one lag ahead $Y(n+1)$, it is assumed the time series is AR(1), that is $p=1$, and that a noninformative prior for the parameters θ and τ; thus, referring to (4.22) and (4.23), it is assumed that $P = 0$, $\alpha = 0$, and $\beta = 0$.
Recall the use of RC 4.4 to generate $n=21$ observations from the AR(1) model (4.15) with $\theta = 0.6$ and $\tau = 1$. Suppose the vector Y consists of the 21 values

generated by RC 4.4, namely Y=(−0.62645381, −0.19222896, −0.95096599, 1.02470121, 0.94432850, −0.25387129, 0.33510628, 0.93938847, 1.13941444, 0.37826027, 1.73873733 1.43308564, 0.23861080, −2.07153341, −0.11798913, −0.11572708, −0.08562651, 0.89246030, 1.35669738, 1.40791975, 1.76372922).

I am assuming the first component of Y is $Y(0) = 0.664$, then it can be shown that $F^* = Y(20) = 1.743, A = \sum_{t=1}^{t=20} Y^2(t-1) = 17.7245, C = \sum_{t=1}^{t=20} Y(t)Y(t-1) =$ 7.5252 and $E = Y^2(20) = 3.0404$ which is sufficient information to compute the mean of the future observation Y(21), given by (4.32). The above can be evaluated by referring to the vector of observations Y, and the student will be asked to compute the average given by formula (4.32).

4.8 Comments and Conclusions

Chapter 4 is the first encounter with the Bayesian analysis of time series. To be considered are the elementary models for time series including white noise, a random walk series, a random walk with drift, and finally the autoregressive class of models. The random walk is first defined followed by the derivation of its first and second moments. The random walk is generalized to include a positive drift parameter. Observations from this model are generated with R. The observations generated are used as data for the Bayesian analysis which is executed with BC 4.1, using 35,000 observations for the simulation starting with 5,000 initial values and the posterior analysis is reported in Table 4.1. One of the most useful time series models is the autoregressive class. This class is defined and the autocorrelation function derived. For the AR(1) model, the posterior distribution is derived employing a noninformative prior for the autocorrelation parameter and the precision of the errors. To illustrate the analysis, R generates 50 observations from an AR(1) model with $\theta = .6$. BC 4.2 is the Bugs code used for the posterior analysis where the results are reported in Table 4.2. The analysis implies the posterior mean of θ is quite close to the value used to generate the data, and Chapter 4 is concluded with another example of an AR(1) time series. The chapter lays the foundation for more complicated time series models. Problem 8 of the exercises introduces another versions of the Bayesian analysis using code based directly on the definition of an AR(1) process, while problem 9 generalizes the Bayesian analysis to AR(2) time series.

  Additional references for Bayesian time series are: Broemeling and Shaarawy,[4,pp. 337–354]which gives an overall picture of Bayesian time series up to the year 1986, Son and Broemeling[5,pp. 337–354] which provides the Bayesian approach to testing for homogeneity of several autoregressive processes, and Broemeling and Son,[6,pp. 927–936] which develops a Bayesian analysis for the classification problem of autoregressive processes.

4.9 Exercises

1. Define a white noise time series and use RC 4.1 to generate 100 observations from such a process with variance $\sigma^2 = 1$.

2. (a) Define the random walk model of (4.2) and derive its mean value function and its variance function.

   (b) Also derive the autocorrelation function.

   (c) With RC 4.2, generate 100 values from the random walk with $\sigma^2 = 1$.

   (d) Use R to plot the 100 values over the interval 1,100.

3. (a) Define the random walk with drift parameter $\zeta = 3$.

   (b) Using RC 1.3, generate 100 observations from a random walk with $\zeta = 3$ and $\sigma^2 = 1$.

   (c) Using the 100 observations generated from the random walk model above, execute a Bayesian analysis with BC 4.1. Execute the simulation with 35,000 observations and 5,000 for the initial run. Use a noninformative prior for μ and σ^2.

   (d) Do your results of the Bayesian analysis? Your answer should agree with Table 4.1.

   (e) What is the posterior mean of the drift parameter?

4. (a) Define the AR(1) class of models (4.9).

   (b) Define stationarity for the AR(1) class of models.

   (c) Give several examples of stationary AR models.

   (d) Using RC (4.4) to generate 100 observations from a AR(1) model with $\theta = .6$ and $\sigma^2 = 1$.

   (e) For the AR(1) model, analytically derive the marginal posterior density of θ assuming the noninformative prior

$$\gamma(\theta, \tau) = 1/\tau, \theta \in R, \tau > 0$$

5. (a) Define an AR(1) process.

   (b) Derive the autocorrelation function of an AR(1) process.

   (c) Based on the 50 observations generated with RC 4.4, perform a Bayesian analysis with BC 4.2.

(d) Verify the results of your analysis with Table 4.2.

6. Refer to Section 4.5, the AR(1) process with $\theta = .9$ and $\sigma^2 = .2$.

    (a) Based on RC 4.5, generate 50 observations from information labeled with the R object data (ar1.s).

    (b) Plot the 50 observations ar1.s.

    (c) Using the data of ar1.s, execute a Bayesian analysis with 45,000 for the MCMC simulation and 5,000 starting values.

    (d) Verify Table 4.3, the posterior analysis.

    (e) What is the posterior mean and 95% credible interval for θ?

7. (a) Show the predictive density of m future observations is given by (4.18).

    (b) Execute a Bayesian analysis with BC 4.3 using 45,000 observations for the simulation and 5,000 starting values.

    (c) The vector X of BC 4.3 is the vector of residuals. What is the posterior mean and 95% credible interval of the residuals. Does the 95% credible interval imply the AR(1) model fits the data? Explain your answer.

8. Consider the AR(1) model of Section 4.4, namely

$$Y(t) = \theta \, Y(t-1) + e(t), t = 1, 2, ..., 51 \tag{4.34}$$

where

$$e(t) \sim nid(0, \tau). \tag{4.35}$$

Recall that for that example $\theta = .6$, $\tau = 1$ and that 50 observations were generated with RC 4.4 and the Bayesian analysis executed with BC 4.2.

The following WinBUGS code is another way to perform the analysis and is different than BC 4.2 in that the multivariate normal is not specifically used for the analysis. Instead, the model (4.34) is included directly in the program statements. The Bayesian analysis is executed with 45,000 observations for the simulation and 5,000 initial values.

BC 4.4

```
model;
{
mu[1]<-theta1*ym1
y[1]~dnorm(mu[1],tau)
for ( t in 2:50){ mu[t]<-theta1*y[t-1]
y[t]~dnorm(mu[t],tau)
}
```

```
# prior distributions
theta1~dnorm(0.0,.001)
tau~dgamma(.01,.01)
ym1~dnorm(0.0,.001)
}
list(y=c(  −0.62645381,  −0.19222896,  −0.95096599,  1.02470121,  0.94432850,
−0.25387129,  0.33510628,  0.93938847,  1.13941444,  0.37826027,  1.73873733,
1.43308564,  0.23861080,  −2.07153341,  −0.11798913,  −0.11572708,  −0.08562651,
0.89246030,  1.35669738,  1.40791975,  1.76372922,  1.84037383,  1.17878928,
−1.28207813, −0.14942113, −0.14578142, −0.24326436, −1.61671100, −1.44817665,
−0.45096443,  1.08810089,  0.55007281,  0.71771530,  0.37682414,  −1.15096507,
−1.10557361,  −1.05763412,  −0.69389387,  0.68368905,  1.17338918,  0.53950991,
0.07034427,  0.73916994,  1.00016516,  −0.08865660,  −0.76068912,  −0.09183151,
0.71343402, 0.31571420, 1.07053625))
list( theta1=.6, tau=1,ym1=22)
```

The posterior analysis is reported in Table 4.5.

See the above code BC 4.3. What are the prior distributions for τ, θ, and y0.?

(a) Compare the posterior mean of θ given by Table 4.5 with that given by Table 4.2.

(b) What factors would account for the difference in the posterior analysis of Table 4.2 and that of Table 4.5.

(c) Portray the posterior density of θ.

(d) Is the posterior density of τ symmetric? Explain your answer.

1. Consider the AR(2) model

$$Y(t) = \theta_1 Y(t-1) + \theta_2 Y(t-2) + e(t) \qquad (4.36)$$

where

$$e(t) \sim nid(0, \tau). \qquad (4.37)$$

TABLE 4.5

Posterior Analysis for AR(1)

Parameter	Value	Mean	SD	Error	2½	Median	97½
θ	.6	.4878	.1353	.000558	.2132	.4906	.7472
τ	1	1.416	.2904	.00103	.9054	1.397	2.038

50 values are generated from (4.36) where $\theta_1 = .31$, $\theta_2 = .32$, and $\tau = 1$ using the following R code:

RC 4.5.

```
set.seed(1)
y<-w<-rnorm(50)
for( t in 3:50) y[t]<-.31*y[t-1]+.32*y[t-2]+w[t]
time <- 1:50
plot(time,y)
```

The time series values appear as the vector y, where

Y=(−0.62645381, 0.18364332, −0.97916440, 1.35050570, 0.43483193, −0.25350866,] 0.54798759, 0.82707809, 1.00753159, 0.27161139, 1.91839081, 1.07146003, 0.32479709, −1.77114558, 0.67981086, −0.40095883, 0.07705197, 0.83941550, 1.10609663, 1.20540424, 1.64660361, 1.67831277, 1.12175510, −1.10454753, 0.63637765, −0.21230688, −0.01796979, −1.54426122, −0.96262137, −0.37463465, 0.93450397, 0.06702541, 0.70749076, 0.18696523, −1.09270329, −0.69390371,−0.95906516, −0.57867278, 0.61373596, 0.76825861, 0.27003208, 0.07619102, 0.80699286, 0.83121211, −0.17284223, −0.49508837, 0.15579505, 0.65840111, 0.14161255, 1.13569597) and a plot of this time series is depicted in Figure 4.3.

Using the data generated by RC 4.5, the Bayesian analysis is executed with BC 4.5 with 150,000 for the simulation and 5,000 initial values. The results of the posterior analysis are shown in Table 4.6.

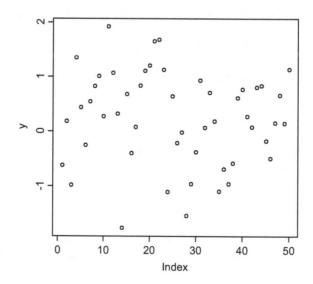

FIGURE 4.3
AR(2) Time Series

TABLE 4.6

Posterior Analysis for AR(2)

Parameter	Value	Mean	SD	Error	2½	Median	97½
τ	1	1.4306	.2933	.001735	.8917	1.387	2.038
θ_1	.32	.2562	.1491	.001795	-.0212	.2547	.5549
θ_2	.33	.1215	.1387	.002405	-.1406	.1169	.404

BC 4.5

```
Model;
{
# AR(2) model
mu[1] <- theta1*ym1+theta2*y0
y[1] ~ dnorm(mu[1],tau)
for(t in 2:50) {
mu[t] <- theta1*y[t-1]+theta2*y[t-1]
y[t] ~ dnorm(mu[t],tau)
}
# Prior distribution
theta1 ~ dnorm(0.0, .001)
theta2~ dnorm(0.01,.001)
tau ~ dgamma(0.01, 0.01)
y0 ~ dnorm(0.0,1.001)
ym1 ~ dnorm(0.0,1.001)
}
# Simulated data: N=50, theta1=.31,theta2=.32, tau=1.00 and y0= 21.34232,
ym1=22.
list(y=c(−0.62645381,  0.18364332,  −0.97916440,  1.35050570,  0.43483193,
−0.25350866,  0.54798759,  0.82707809,  1.00753159,  0.27161139,  1.91839081,
1.07146003,  0.32479709,  −1.77114558,  0.67981086,  −0.40095883,  0.07705197,
0.83941550,  1.10609663,  1.20540424,  1.64660361,  1.67831277,  1.12175510,
−1.10454753, 0.63637765,  −0.21230688,  −0.01796979,  −1.54426122,  −0.96262137,
−0.37463465,  0.93450397,  0.06702541,  0.70749076,  0.18696523,  −1.09270329,
−0.69390371,  −0.95906516,  −0.57867278,  0.61373596,  0.76825861,  0.27003208,
0.07619102,  0.80699286,  0.83121211,  −0.17284223,  −0.49508837,  0.15579505,
0.65840111, 0.14161255, 1.13569597))
# Initial values
list(theta1=.31,theta2=.32, tau=1.0,y0=20.0,ym1=20)
```

The posterior mean of θ_2 is quite close to its true value of .33; however, the posterior mean of θ_1 is not that close to its true value, but the 95% credible interval does include the true value of .32.

(a) Using RC 4.5, generate 100 observations from the model (4.36)

(b) Perform a Bayesian analysis using BC 4.5 with 250,000 observations for the simulation and 5,000 initial values.

(c) Compare the results of your analysis with that reported in Table 4.6.

(d) Based on your posterior analysis, does yours give more accurate results compared to those in Table 4.6.

(e) Test the hypothesis H: $\theta_1 = .31$. This tests the hypothesis that the value of θ_1 used to generate the data is indeed .31.

(f) Does your posterior analysis imply the AR(2) model is stationary? Explain why?

11.(a) For an AR(2) model (4.36) and (4.37), what is the autocorrelation function?

According Cryer and Chan[2, p.73], the autocovariance function satisfies the recursive relation

$$\gamma_k = \theta_1 \gamma_{k-1} + \theta_2 \gamma_{k-2}, k = 1, 2, \ldots \qquad (4.38)$$

where γ_k is the lag k autocovariance; thus, it follows that the autocorrelation satisfies the recursive relation

$$\rho_k = \theta_1 \rho_{k-1} + \theta_2 \rho_{k-2}, k = 1, 2, \ldots, \qquad (4.39)$$

where

$$\rho_k = \theta_1 \rho_{k-1} + \theta_2 \rho_{k-2}, k = 1, 2, \ldots$$
$$\rho_k = \gamma_k / \gamma_0 \qquad (4.40)$$
$$\gamma_0 = [(1 - \theta_2)/(1 + \theta_2)]\sigma^2/[(1 - \theta_2)^2 - \theta_1^2]$$

and σ^2 is the variance of residuals (4.37).

(c) Derive the above equations (4.38)–(4.40).

12. Refer to Section 4.7 on the Bayesian prediction.

(a) Derive the predictive density (4.31).

(b) Derive the mean of the future observation $Y(n+1)$ given by (4.32).

(c) (c) Based on the 21 observations generated by RC 4.4 for the AR(1) series with parameters $\theta = 0.6$ and $\tau = 1$, compute the mean of the future observation Y(21). There is sufficient information to do the calculation.

References

1. Cowpertwait, P.S.P. and Metcalfe, A.V. (2009). *Introductory Time Series with R.* Springer, New York.
2. Cryer, J.D. and Chan, K.S. (2008). *Time Series Analysis: with Applications in R.* Springer, New York.
3. Broemeling, L.D. (1985). *Bayesian Analysis of Linear Models.* Marcel-Dekker Inc., New York.
4. Broemeling, L.D. and Shaarawy, S. (1986). "Bayesian Analysis of Time Series," in *Chapter 21 of Bayesian Inference and Decision Techniques*, edited by P. Goel and A. Zellner, Elesivier Science Publishers. Baltimore, MD.
5. Son, M.S. and Broemeling, L.D. (1990). "On testing for homogeneity of several autoregressive processes", *Statistica*, 50(2), 239–245.
6. Broemeling, L.D. and Son, M.S. (1982). "The classification problem with auto-regressive processes", *Communications in Statistics, Theory and Methods*, Vol. 16 (4), 927–936.

5

Times Series and Regression

5.1 Introduction

In this section, various regression models are introduced. First to be considered is the estimation of the trend in linear models with autocorrelated errors, then the next part is to provide inferences for the seasonal effects using harmonic and latent variables. Time series regression models are different from the usual regression models in that the errors are autocorrelated. One of the first linear models to be studied is simple linear regression following an AR(1) process. What is the effect of autocorrelation on the usual estimates of the regression coefficients? If the correlation is positive, the estimated standard errors of the estimates tend to be less than the estimated standard errors of the estimates assuming no correlation. Of course, a corresponding scenario holds for the Bayesian estimates of the regression coefficients.

5.2 Linear Models

A model is linear if

$$Y(t) = \beta_0 + \beta_1 X_1(t) + \beta_2 X_2(t) + \ldots + \beta_m X_m(t) + Z(t) \qquad (5.1)$$

where $Y(t)$ is the observation of the dependent variable at time t, $X_i(t)$ is the observation of the i-th independent variable at time t, and finally $Z(t)$ is the error term at time t. The errors $Z(t), t = 1, 2, \ldots, n$ are assumed to have mean zero, have a constant variance, and are autocorrelated. Our goal is to compute Bayesian inferences for the m regression coefficients β_i and the unknown parameters of the error process.

As a first example, consider the linear regression model

$$Y(t) = 50 + 3t + Z(t), t = 1, 2, \ldots, 100 \tag{5.2}$$

where the $Z(t)$ follow an AR(1) process with autocorrelation θ.

Consider RC 9.1 which will generate 100 observations from the linear regression model. The model is assumed to have a standard deviation of 10 for the Gaussian white noise of the AR(1) process and the autocorrelation coefficient is assigned the value $\theta = .6$, while the regression coefficients are assumed to be 3 for the slope and 50 for the intercept. The first fifty values generated by RC 5.1 appear below the code. The fifty values of the dependent variable are the components of vector Y.

RC 5.1
```
> set.seed(1)
> u<-w<-rnorm(100,sd=10)
> for ( t in 2:100) u[t]<-.6*u[t-1]+w[t]
> time<-1:100
> y<-50+3*time+u
> plot(time, y)
```
Y=(46.73546,54.07771,49.49034,72.24701,74.44328, 65.46129,74.35106,83.39388, 88.39414,83.78260, 100.38737, 100.33086,91.38611,71.28467,93.82011,96.84273, 100.14373, 112.92460, 120.56697, 124.07920, 130.63729, 134.40374, 130.78789, 109.17922, 123.50579,126.54219, 128.56736, 117.83289, 122.51823, 135.49036, 153.88101, 151.50073, 156.17715, 155.76824, 143.49035, 146.94426, 150.42366, 157.06106, 173.83689, 181.73389, 178.39510, 176.70344,186.39170, 192.00165, 184.11343, 180.39311, 190.08168, 201.13434, 200.15714, 210.70536)

A plot of the simple linear regression dependent variables versus time over fifty days appears in Figure 5.1, and the linear trend is obvious. What is the posterior mean of the intercept, slope, and autocorrelation?

It appears the observation begin at 50 starting at time 0.

A goal of the Bayesian analysis is to estimate the regression coefficient and the autocorrelation θ. In the statements of **BC 5.1,** beta0 is the intercept and beta1 the slope of the regression model with autocorrelation theta. The 20 observations is a vector with a multivariate normal distribution with mean vector consisting of 20 values of $50+3t$, $t=1,2, \ldots ,20$, and a 20-by-20 precision matrix which is the inverse of the variance covariance matrix of the variance–covariance matrix of an AR(1) error process.

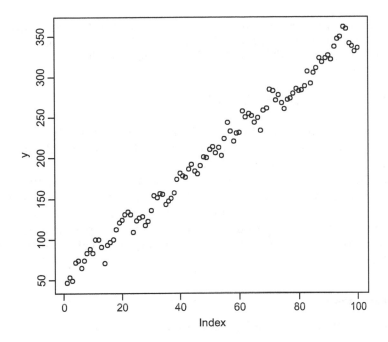

FIGURE 5.1
Simple Linear Regression

BC 5.1
model;
{

```
v~dgamma(.1,.01)
theta~dbeta(6,4)
beta0~dnorm(0,.001)
beta1~dnorm(0,.001)

    Y[1,1:20] ~ dmnorm(mu[], tau[, ])


    for( i in 1:20){mu[i]<-beta0+beta1*i}


    tau[1:20,1:20]<-inverse(Sigma[,])
    for(i in 1:20){Sigma[i,i]<-v/(1-theta*theta)}
    for(i in 1:20){for(j in i+1:20){Sigma[i,j]<- v*pow(theta,j)*1/(1-theta*theta)}}
    for( i in 2:20){ for(j in 1: i-1){Sigma[i,j]<-v*pow(theta,i-1)*1/(1-theta*theta)}}


    }
```

list(Y= structure(.Data=c(46.73546,54.07771,49.49034,72.24701,74.44328,
65.46129, 74.35106,83.39388,88.39414,83.78260, 100.38737, 100.33086,91.38611,
71.28467,93.82011,96.84273, 100.14373, 112.92460, 120.56697, 124.07920),.
Dim=c(1,20)))

list(theta=.6,v=100, beta0=50, beta1=3)

The posterior analysis for the AR(1) regression is executed by BC 5.1 using
45,000 observations for the simulation and 5,000 for the initial run, and the
results of that analysis appears in Table 5.1.

Recall the true value of the intercept is 50 and the corresponding mean is
47.99 which is in good agreement, while the true value for the slope is 3,
which is estimated as 3.483 with the posterior mean. The 95% credible
intervals for the four parameters imply that the posterior means (or
medians) are quite close to the corresponding 'true' values in the MCMC
simulation.

5.3 Linear Regression with Seasonal Effects and Autoregressive Models

Our next example is taken from Cowpertwiat and Metcalfe[1, pp. 101–105] and
concerns a model with trend and seasonal effect where the data is based
on the model

$$Y(t) = 0.1 + .005t + .001t^2 + \sin(2\pi t/12) + .2\sin(4\pi t/12)$$
$$+ .1\sin(8\pi t/12) + .1\cos(8\pi t/12) + W(t) \tag{5.3}$$

where the $W(t)$ are autocorrelated with coefficient $\theta = .6$ and $\sigma^2 = .25$, and
$n = 120$. Note that the model is linear model with a quadratic trend and
seasonal effects represented by sinusoidal waves with very small amplitudes and frequencies of 1, 2, and 3 cycles per unit time. RC 5.2 generates

TABLE 5.1

Posterior Analysis for Simple Linear Regression

Parameter	Value	Mean	SD	Error	2½½	Median	97½
β_0	50	47.99	6.828	.1323	32.9	48.35	60.57
β_1	3	3.483	.4805	.009213	2.567	3.47	4.499
θ	.6	.6509	.1487	.006511	.3352	.6653	.8933
σ^2	100	54.51	24.33	.2522	20.92	50.01	114

120 values for the dependent variable given by (9.46) and these are in the list statement of the WinBUGS program BC 5.2 below.

RC 5.2

```
set.seed(1)
time<-1:(10*12)
w<-rnorm(10*12, sd=.5)
Trend<-0.1+.005*time+.001*time^2
Seasonal<- sin(2*pi*time/12)+0.2*sin(2*pi*2*time/12)+
0.1*sin(2*pi*4*time/12)+0.1*cos(2*pi*4*time/12)
x<-Trend+Seasonal+w
```

A plot of the dependent variable over time is depicted in Figure 5.2. The general version of the specific model (5.2) is defined as

$$
\begin{aligned}
Y(t) = &\beta_0 + \beta_1 t + \beta_2 t^2 + \beta_3 t^3 + \beta_4 \sin(2\pi t/12) \\
&\beta_5 \sin(8\pi t/12) + \beta_6 \cos(8\pi t/12) + W(t)
\end{aligned}
\tag{5.4}
$$

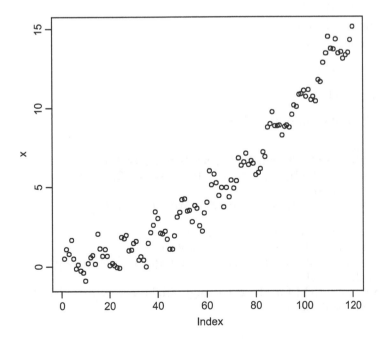

FIGURE 5.2
Regression with Quadratic Trend and Seasonal Effects

where the $\beta_i, i = 0, 1, \ldots 6$ are unknown regression coefficients, and $W(t)$ a sequence of errors which are correlated. The Bayesian analysis is based on the 30 values of the dependent variable of model (9.46) and the goal is to estimate the $\beta_i, i = 0, 1, \ldots, 6$, the autocorrelation coefficient θ, and variance σ^2. BC 5.2 is executed with 45,000 observations for the simulation and 5,000 for the burn-in. Note that the vector of 30 observations is specified as a multivariate normal distribution with 30 by 1 mean vector (9.46) and precision matrix τ which is the inverse of the variance covariance matrix of an AR(1) process with correlation θ. The parameters of the model are given noninformative prior distributions. See BC 5.2 for the specifications.

The Bayesian analysis is executed with 45,000 observations for the simulation with 5,000 starting values.

BC 5.2

```
model;
{
beta0~dnorm(0,.001)
beta1~dnorm(0,.001)
beta2~dnorm(0,.001)
beta3~dnorm(0,.001)
beta4~dnorm(0,.001)
beta5~dnorm(0,.001)
beta6~dnorm(0,.001)

theta~dbeta(6,4)

v~dgamma(.001,.001)

Y[1,1:30]~dmnorm(mu[],tau[,])

for( t in 1:30){
mu[t]<-beta0+beta1*t+beta2*t*t+beta3*sin(2*3.1416*t/12)
+beta4*sin(4*3.1416*t/12)+beta5*sin(8*3.1416*t/12)+
beta6*cos(8*3.1416*t/12)}
#Sigma is the variance covariance matrix of an AR(1)

        tau[1:30,1:30]<-inverse(Sigma[,])
        for(i in 1:30){Sigma[i,i]<-v/(1-theta*theta)}
        for(i in 1:30){for(j in i+1:30){Sigma[i,j]<- v*pow(theta,j)*1/
        (1-theta*theta)}}
        for( i in 2:30){ for(j in 1: i-1){Sigma[i,j]<-v*pow(theta,i-1)*1/
        (1-theta*theta)}}


}
```

list(Y =structure(.Data=c(0.50258072,1.10844961,0.80618569,1.66306326,
0.50494626, −0.14423419,0.13752215, −0.25626051, −0.38610932, −0.90532214,
0.22208296,0.59892162,0.73318733,0.16127800,2.06246546,1.14295606,0.65609725,
1.08591811,0.67641822,0.06752780,0.20548869,0.08244021, −0.02852513,
−0.09867585, 1.86972050,1.78056357,1.98610225,.01804667,1.03711735,
1.45897078),.Dim=c(1,30)))
list(theta=.6,v=.25,beta0=.1,beta1=.1,beta2=.2,beta3=.2,beta4=.2,beta5=.2,
beta6=.1)

The posterior analysis for the model with quadratic trend and seasonal effects is revealed in Table 5.2.

The values used to generate the data are listed in the second column and should be compared to their corresponding posterior means. For example, the value of β_3 used for the simulation is 1 compared to its posterior mean of .7606 and 95% credible interval of (.4304,1.09) which indeed includes 1! The posterior mean for σ^2 is very close to its nominal value of .25 used to generate the data and its 95% credible interval does indeed include .25!

5.4 Bayesian Inference for a Nonlinear Trend in Time Series

As we have seen linear model has a wide range of applications, but so do time series that have a nonlinear trend and autocorrelated errors. Consider the process

$$Y(t) = \exp(\beta_0 + \beta_1 t) + Z(t) \tag{5.5}$$

TABLE 5.2

Posterior Analysis for Seasonal Effects

Parameter	Value	Mean	SD	Error	2½	Median	97½
β_0	1	.086	.459	.02289	−.8377	.0971	.9962
β_1	.005	.0345	.06122	.00341	−.0875	.0321	.159
β_2	.001	.0003	.0018	.00010	−.004	−.0002	.00332
β_3	1	.7606	.1669	.00295	.4304	.7601	1.09
β_4	.2	.1711	.1525	.00094	−.1315	.1722	.4722
β_5	.1	.0453	.1468	.00092	−.2424	.0454	.337
β_6	.1	.1092	.nb5	.00099	−.186	.1084	.4041
θ	.6	.6012	.1461	.00189	.3023	.6077	.8599
σ^2	.25	.2139	.0901	.00100	.0849	.1991	.428

where β_0 and β_1 are unknown parameters and the residuals $Z(t)$ on the log scale form an AR(1) process with autocorrelation θ. The following R code is taken from Cowpertwait and Metcalfe[1, pp.113–114] and generates 100 observations from the nonlinear model with $\beta_0 = 1$, $\beta_1 = .05$, $\theta = .6$, and $\sigma = 2$.

RC 5.3.

```
set.seed(1)
w<-rnorm(100,sd =2)
z<-rep(0,100)
for ( t in 2:100) z[t]<-−0.6*z[t-1]+w[t]
Time<-−1:100
f<- function(x) exp(1+0.05*x)
x<-f(Time)+z
```

The first 50 values of the simulation RC 5.3 appear as components of the vector y.

y=(2.857651, 3.371453, 1.707308, 5.640147, 5.541377, 3.258980, 4.586094, 5.969050, 6.562987, 5.250836, 8.196521, 7.823749,5.686929, 1.332517,5.519606, 5.818782,6.188920,8.471027, 9.742210,10.204972,11.295405,11.846945,10.942453, 6.460867,9.188900,9.682623,9.999043,7.789756,8.691994,11.280566,14.983306, 14.563884,15.589469,15.633380,13.340702,14.233500, 15.172514,16.786358, 20.473332,22.432315,22.194364,22.338640, 24.814404,26.532861,25.613027, 25.591261,28.319071,31.390968, 32.131821,35.256524)

The Bayesian analysis is based on 30 observations generated by RC 9.10 and will focus on the estimation of the unknown parameters β_0, β_1, θ, and σ^2, and is executed with **BC 5.3** using 35,000 observations for the simulation and 5,000 for the burn-in.

The 30 observations are the components of a vector Y which has a multivariate normal distribution with mean vector given by (9.48) and precision matrix which is the inverse of the variance covariance matrix of an AR(1) process with correlation θ.

BC 5.3

```
model;
{
beta0~dnorm(0,.001)
beta1~dnorm(0,.001)
theta~dbeta(6,4)
v~dgamma(.001,.001)
```

TABLE 5.3

Posterior Analysis for Nonlinear Model

Parameter	Value	Mean	SD	Error	2½	Median	97½
β_0	1	1.407	.1794	.00945	1.021	1.41	1.757
β_1	.05	.0327	.0077	.00040	.0170	.0328	.0484
θ	.6	.6349	.1513	.00269	.3218	.6455	.891
σ^2	4	2.449	1.056	.01645	.8932	2.303	4.932

```
for( t in 1:30){mu[t]<-exp(beta0+beta1*t)}
Y[1,1:30]~dmnorm(mu[],tau[,])

    tau[1:30,1:30]<-inverse(Sigma[,])
        for(i in 1:30){Sigma[i,i]<-v/(1-theta*theta)}
        for(i in 1:30){for(j in i+1:30){Sigma[i,j]<- v*pow(theta,j)*1/
        (1-theta*theta)}}
        for( i in 2:30){ for(j in 1: i-1){Sigma[i,j]<-v*pow(theta,i-1)*1/
        (1-theta*theta)}}

}

list(
Y=structure(.Data=c(2.857651,3.371453,1.707308,5.640147,5.541377,3.258980,
4.586094,5.969050,6.562987,5.250836,8.196521,7.823749,5.686929,1.332517,
5.519606,5.818782,6.188920,8.471027,
9.742210,10.204972,11.295405,11.846945,10.942453,6.460867,9.188900,
9.682623,9.999043,7.789756,8.691994,11.280566),.Dim=c(1,30)))

list(beta0=1,beta1=.05,theta=. 6, v=4)
```

The posterior analysis is reported in Table 5.3

The posterior analysis reveals that the posterior mean of .6349 for θ is very close to the value used to generate the observations, and the same holds for the other parameters, as for example σ^2, with a posterior mean of 2.449, which is also very close to the value used to generate the observations.

5.5 Nonlinear Trend with Seasonal Effects

This is the most complex of the regression models to be studied in this chapter. Consider the following nonlinear model with exponential trend, seasonal effects, and AR(1) errors. Recall previously considered is a linear model (5.4) with AR(1) errors and a nonlinear model (5.5) with AR(1) errors. Let

$$Y(t) = \beta_0 + \beta_1 \exp(-\beta_2 t) + \beta_3 \sin(2\pi t/12) + \beta_4 \sin(8\pi t/12) + \beta_5 \cos(8\pi t/12) + u(t)$$
$$u(t) = \theta u(t-1) + w(t)$$

$$(5.6)$$

where $w(t)$ is white noise with variance 1, and $u(t)$ is an autoregressive process with correlation θ and variance σ^2, while the $\beta_i, i = 0, 1, \ldots, 5$ are unknown regression coefficients. The goal is to estimate the parameters and test hypotheses about the parameters of this model with the Bayesian approach. The following R code generates 30 observations from the above model (5.6) and the code statements closely follow the symbols in the model (5.6)

RC 5.4
```
set.seed(1)
time<-1:30
u<-w<-rnorm(30, sd= 1)
Trend<-3+exp(-2*time)
Seasonal<-1*sin(2*pi*time/12)+2*sin(8*pi*time/12)+
2*cos(8*pi*time/12)
for(t in 2:30) u[t]<-.6*u[t-1]+w[t]
x<-Trend+Seasonal+u
```

Note to generate the observations with RC 4.5, the parameter values are: $\beta_0 = 3, \beta_1 = 1, \beta_2 = 2, \beta_3 = 1, \beta_4 = 2, \beta_5 = 2, \theta = .6$

The 30 x values are x=(3.7409323,0.9600613,5.0515128, 5.6231129, 1.7123231, 4.7461349, 3.5671579,0.3413124, 5.1394145,3.2442857,1.5066865, 6.4330856, 4.4706616, −0.9375588, 5.8820109,4.4823491,0.6823227, 5.8924603, 4.5887482, 0.8098435, 5.7637292,4.7063992, 0.9467385, 3.7179219, 4.0826297, 0.9881932, 5.7567356,2.9813652, −0.6802275,4.5490356),

and the time series graph is portrayed in Figure 5.3.

It is obvious that the exponential trend cannot be identified from the plot because the seasonal effects are masking the trend.

Using these 30 values as the data, the Bayesian analysis is executed with 25,000 observations for the simulation with 3,000 for the initial values. See BC 5.4 for the WinBUGS code for the MCMC simulation. Note that the prior distributions for the regression coefficients are normal centered at the true mean with variance 1. This is quite informative prior information compared to the usual improper priors that could have been employed. The student will be asked to repeat the analysis but with noninformative

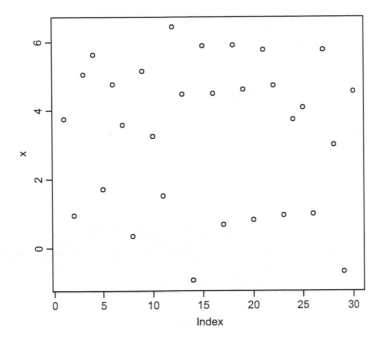

FIGURE 5.3
Nonlinear Regression with Exponential Trend and Seasonal Effects

normal prior distributions with mean 0 and precision .001 (variance 1000). In this case, one would expect the posterior means to be not as accurate compared to using the more informative prior of BC 5.4. This will clearly show the effect of prior information on the posterior distribution.

BC 5.4.

```
model;
{
beta0~dnorm(3,1)
beta1~dnorm(1,1)
theta~dbeta(6,4)
beta2~dnorm(2,1)
beta3~dnorm(1,1)
beta4~dnorm(2,1)
beta5~dnorm(2,1)
v~dgamma(.1,.1)
pi<-2.416
```

```
for( t in 1:30){mu[t]<-beta0+beta1*exp(-beta2*t)+beta3*sin(2*pi*t/12)+
beta4*sin(8*pi*t/12)+beta5*cos(8*pi*t/12)}
Y[1,1:30]~dmnorm(mu[],tau[,])

  tau[1:30,1:30]<-inverse(Sigma[,])
      for(i in 1:30){Sigma[i,i]<-v/(1-theta*theta)}
      for(i in 1:30){for(j in i+1:30){Sigma[i,j]<- v*pow(theta,j)*
      1/(1-theta*theta)}}
      for( i in 2:30){ for(j in 1: i-1){Sigma[i,j]<-v*pow(theta,i-1)*
      1/(1-theta*theta)}}

}

list(
Y=structure(.Data=c( 3.7409323,0.9600613,5.0515128,5.6231129,1.7123231,
4.7461349, 3.5671579,0.3413124,5.1394145,3.2442857,1.5066865,6.4330856,
4.4706616, −0.9375588,5.8820109,4.4823491,0.6823227,5.8924603, 4.5887482,
0.8098435,5.7637292,4.7063992,0.9467385,3.7179219, 4.0826297,0.9881932,
5.7567356,2.9813652, −0.6802275,4.5490356),.
Dim=c(1,30)))

list(beta0=3, beta1=1, theta=.6, v=1,beta2=2, beta3=1, beta4=2, beta5=2)
```

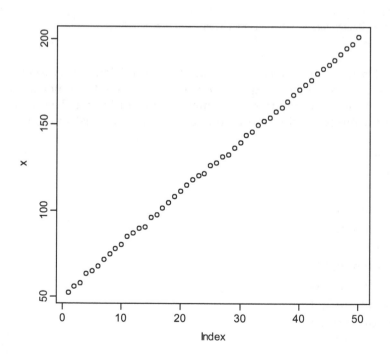

FIGURE 5.4
The Simple Linear Regression with AR(2) Errors

TABLE 5.4

Posterior Analysis for Nonlinear Model with Seasonal Effects

Parameter	Value	Mean	SD	Error	2½	Median	97½
β_0	3	3.331	.4224	.0028	2.489	3.335	4.16
β_1	1	.9865	.9909	.0066	−.941	.9826	2.92
β_2	2	2.055	.9381	.0103	.3618	2.031	3.969
β_3	1	.5018	.5222	.0035	−.518	.4932	1.54
β_4	2	.3367	.5147	.0031	−.653	.3241	1.378
β_5	2	.9738	.518	.0035	−.0480	.9723	2.005
θ	.6	.5894	.1407	.0016	.2991	.5982	.8347
v	1	3.467	1.034	.0515	.525	3.268	6.576

The results of the Bayesian analysis are depicted with Figure 5.4 (Table 5.4). One observes that the Bayesian analysis provides somewhat accurate estimates (via the posterior mean and median) of the various parameters of this model. For example, the posterior mean of β_0 is 3.331 which compares to the 'true' value of 3. Also, the posterior mean of θ is .5894 compared to its true value of .6 and the 95% credible interval is (.2991,.8347).

5.6 Regression with AR(2) Errors

Recall problem 9 of Chapter 4, where the Bayesian analysis of the AR(2) model is introduced. This section will describe a Bayesian analysis for regression models with AR(2) that includes those of the previous section including the simple linear, a model that includes quadratic effects, a nonlinear model that includes exponential trends, and finally a nonlinear model with seasonal effects.

5.6.1 Simple Linear Regression Model

RC 5.5

Consider the following simple linear regression model with AR(2) errors.

$$y(t) = \beta_0 + \beta_1 t + w(t)$$
$$\text{and} \tag{5.7}$$
$$w(t) = \theta_1 w(t-1) + \theta_2 w(t-2) + e(t)$$

where

the e(t)~ nid(0, τ). The unknown parameters are $\beta_0, \beta_1, \theta_1, \theta_2$, and τ.

$$t = 1, 2, \ldots, n.$$

In order to demonstrate the Bayesian approach, the parameters are assigned the following values: $\beta_0 = 50, \beta_1 = 3, \theta_1 = .2, \theta_2 = -.70$, and $\tau = 1$. Recall that an AR(2) model is stationary if $\theta_1 + \theta_2 < 1, \theta_2 - \theta_1 < 1$, and $|\theta_2| < 1$.

The following R code generates 50 observations from the (5.7) with the above parameter values.

RC 5.5

```
set.seed(1)
time<-1:50
u<-w<-rnorm(50, sd= 1)
Trend<-50+3*time
for(t in 3:50) u[t]<-.2*u[t-1]-.7*u[t-2]+w[t]
x<-Trend+u
```

The following 50 values generated are denoted by the vector x and appear in Figure 5.4.

x=(52.37355, 56.18364,58.63962,63.39465,65.86071,66.37542,70.56002, 75.78754,78.24128,78.69159,83.38121,87.38197,88.38831,88.69558, 95.89223, 100.44660, 100.84857, 103.20093, 107.76741, 111.30673, 113.64314, 115.99605, 118.62358, 119.93813, 125.47095, 129.48137, 130.81082, 131.45445, 136.14517, 142.02886, 145.36283, 144.94958, 147.52360, 152.38621, 154.73366, 157.26139, 160.64443, 164.38660, 168.42625, 170.77781, 171.99266, 175.00071, 180.20224, 183.49662, 183.76900, 185.99867, 191.82602, 196.33466, 196.77637, 199.20212)

The Bayesian analysis is executed with BC 5.5 using 525,000 observations with 5,000 initial values, and the posterior analysis is reported in Table 5.5.

BC 5.5

```
Model;
{
# AR(2) model
mu[1] <-beta0+beta1+theta1*y0+theta2*ym1
y[1]~dnorm(mu[1],tau)
for(t in 3:50) {
mu[t] <- beta0+beta1*t+ theta1*y[t-1]+theta2*y[t-2]
y[t]~ dnorm(mu[t],tau)
}
```

TABLE 5.5

Simple Linear Regression with AR(2) Errors

Parameter	Value	Mean	SD	Error	2½	Median	97½
β_0	50	50	1.211	.03447	47.59	50.02	52.35
β_1	3	3.088	.0876	.0026	2.916	3.0989	3.26
θ_1	.2	.4615	.0775	.0028	.2928	.4615	.6062
θ_2	−.7	−.4911	.0836	.0030	−.6426	−.4916	−.3126
τ	1	.9404	.1989	.0011	.5927	.9262	1.369
ρ_1	.117	.3049	.0356	.00155	.2144	.307	.3646

```
# Prior distribution
beta0~dnorm(0,.001)
beta1~dnorm(0,.001)
theta1~ dnorm(.2, .1)
theta2~ dnorm(−.7,.1)
tau~ dgamma(0.01, 0.01)
y0~ dnorm(0.0,1.001)
ym1 ~ dnorm(0.0,1.001)
}

# Simulated data: N=50, theta1=.2,theta2=−.7, tau=1.00 and y0= 21.34232,
ym1=22.
list(y=c(52.37355, 56.18364,58.63962,63.39465,65.86071,66.37542,70.56002,
75.78754,78.24128,78.69159,83.38121,87.38197,88.38831,88.69558,
95.89223, 100.44660, 100.84857, 103.20093, 107.76741, 111.30673, 113.64314,
115.99605, 118.62358, 119.93813, 125.47095, 129.48137, 130.81082, 131.45445,
136.14517, 142.02886, 145.36283, 144.94958, 147.52360, 152.38621, 154.73366,
157.26139, 160.64443, 164.38660, 168.42625, 170.77781, 171.99266, 175.00071,
180.20224, 183.49662, 183.76900, 185.99867, 191.82602, 196.33466, 196.77637,
199.20212))

# Initial values
list(theta1=.2,theta2=−.70, tau=1.0,y0=48,ym1=49,beta0=50,beta1=3)
```

The Bayesian analysis provides excellent estimates for β_0, β_1, and τ; however, this is not true for the parameters θ_1, θ_2 of the autoregressive process. It is usually much more difficult to estimate accurately parameters of an autoregressive process using WinBUGS. Also, note the 'true' value of the lag one autocorrelation is .117 compared to the estimated value of .3049 via the posterior mean.

5.7 Comments and Conclusions

The chapter begins with the definition of a linear regression model with autoregressive errors. For all regression models, the errors follow an AR(1) model. A simple linear regression model with AR(1) errors is the next topic of the chapter. For each regression model, the data for the model with known parameters is generated with R code; then using that data as sample information, a Bayesian analysis is performed that includes computing the posterior distribution for each parameter of the model. WinBUGS code is always used for the posterior analysis which is based on an MCMC simulation with typically 35,000 simulations beginning with 5,000 for the initial group. The following regression model with AR(1) errors are studied: (1) simple linear regression, (2) a regression with quadratic over time plus seasonal effects, (3) nonlinear trend using the exponential function, and (4) nonlinear trend (using the exponential) plus seasonal effects. It should be noted that the student will be asked to develop a Bayesian analysis for the above regression models, but with an AR(2) process for the errors. It will be a challenge to write the WinBUGS code for AR(2) error process.

Additional references that are relevant to this chapter are Cryer and Chan,[2, ch. 11] Brillinger,[3] Box and Jenkins,[4] and Bloomfield.[5]

5.8 Exercises

(1). Write a two-page essay about Chapter 5. Describe the basic concepts introduced in this chapter. This includes a portrayal of the role that R plays in the fitting of the various regression models. Also, provide a characterization of the four types of regression models appearing in the chapter and explain how WinBUGS is used to perform the Bayesian analysis. The models used AR(1) processes for the errors of the series. Instead if one had used independent identically distributed random variables for the errors, how would the Bayesian analysis differ from what was accomplished using AR(1) processes for the errors?

(2). Define carefully the linear model defined by (5.1).

(3). Consider the simple linear regression model with intercept $\beta_0 = 50$ and slope $\beta_1 = 3$ and AR(1) errors with parameters $\theta = .6$ and $\sigma^2 = 1$.

    (a). With RC 5.1, generate 100 observations from the regression model defined above.

(b). Plot the 100 values generated with RC 5.1. What R command did you use for the plot?

(c). Use BC 5.1 to execute the Bayesian analysis using the first 20 values generated with RC 5.1. Assume noninformative normal priors for the regression coefficients with mean 0 and precision .001. For the correlation coefficient θ, assume a beta (6,4) prior distribution and for the precision parameter $\tau = 1/\sigma^2$ assume the noninformative gamma (.001,.001) prior.

(d). Execute the posterior analysis with BC 5.1 using 35,000 observations for the simulation and 5,000 starting observations.

(e). Verify Table 5.1, the posterior analysis for the simple linear regression model.

(f). What is the posterior mean and 95% credible interval of β_0? Is the posterior mean close to its 'true' value of 50?

(4). Consider the regression model (5.3) with quadratic trend, seasonal effects, plus AR(1) errors. Let the parameters be:
$\beta_0 = .1, \beta_1 = .005, \beta_2 = .001, \beta_3 = 1, \beta_4 = .2, \beta_5 = .1, \beta_6 = .1$, and $\theta = .6$.

(a). Using RC 5.2, generate 120 observations from this process with the above parameter values.

(b). Plot the 120 values over time. What R command did you use for the plot? Does the plot appear similar to Figure 5.2?

(c). Execute the Bayesian analysis with BC 5.2 using 45,000 observations for the simulation and 5,000 starting values. Use the first 30 observations generated by RC 5.2.

(d). Compare you results to those portrayed in Table 5.2.

(e). What are the posterior means for the regression coefficients. What is the 95% credible interval for θ.

(f). Are the posterior means close to their 'true' values. Explain your answer.

(5). Consider the nonlinear model (5.5) with parameter values: $\theta = .6$ $\beta_0 = 1, \beta_1 = .05$ and $\sigma = 2$.

(a). Use RC 5.3 to generate 30 observations from the model (5.5) and the parameters specified above.

(b). With the 30 observations generated by RC 5.3, execute a Bayesian analysis with BC 5.3 using 35,000 observations for the simulation and 5,000 starting values.

(c). Compare your posterior analysis with that reported in Table 5.3. Your results should be similar to those reported in the table.

(d). What is the posterior mean of θ? What is its 95% credible interval?

(e). Give an estimate of the lag 2 correlation.

(6). Refer to Section 5.5.

(a). Describe the regression model (5.6).

(b). Let the parameters of this process be: $\theta = .6, \beta_0 = 3$, $\beta_1 = 1, \sigma = 2$, $\beta_2 = 2$, $\beta_3 = 1, \beta_4 = 2, \beta_5 = 2$, and $\sigma^2 = 1$. Based on these parameter values, use RC 5.4 to generate 30 observations from the regression model (5.6).

(c). With R, plot the 30 values generated with RC 5.4. Compare your plot with Figure 5.3.

(d). With the 30 values, execute the Bayesian analysis with 30,000 observations for the MCMC simulation and 5,000 starting values.

(e). Are your posterior results similar to Table 5.4?

(f). Refer to Table 5.4; what posterior mean is closest to the corresponding 'true' value?

(7). Cowpertwait and Metcalfe[1, p. 4] discuss the number of international airline booking over the period 1949–1960, which are shown below: in Table 5.6. Consider the plot execute with the R statement plot(Air Passengers) displayed in Figure 5.5.

(a). The trend appears to be increasing in a linear or quadratic fashion with obvious seasonal effects.

(b). Propose a time series regression model that will fit this series. Use AR(1) for the errors of the model.

(c). Describe the Bayesian analysis that will estimate the parameters of the model.

TABLE 5.6

International Airline Bookings

	Jan	Feb	Mar	Apr	May	Jun	Jul	Aug	Sep	Oct	Nov	Dec
1949	112	118	132	129	121	135	148	148	136	119	104	118
1950	115	126	141	135	125	149	170	170	158	133	114	140
1951	145	150	178	163	172	178	199	199	184	162	146	166
1952	171	180	193	181	183	218	230	242	209	191	172	194
1953	196	196	236	235	229	243	264	272	237	211	180	201
1954	204	188	235	227	234	264	302	293	259	229	203	229
1955	242	233	267	269	270	315	364	347	312	274	237	278
1956	284	277	317	313	318	374	413	405	355	306	271	306
1957	315	301	356	348	355	422	465	467	404	347	305	336
1958	340	318	362	348	363	435	491	505	404	359	310	337
1959	360	342	406	396	420	472	548	559	463	407	362	405
1960	417	391	419	461	472	535	622	606	508	461	390	

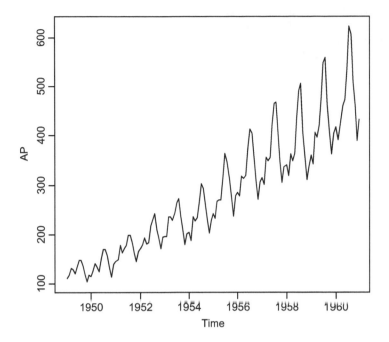

FIGURE 5.5
International Airline Passenger Bookings

(d). Use WinBUGS for the posterior analysis. Report the analysis with a table that consists of a row for each parameter of the model. Each row should report the posterior mean, posterior standard deviation, the posterior 2½ percentile, the posterior median, and the Posterior 97½ percentile.

(e). Explain how your model fits the data.

(8). Refer to the regression model (5.2), the simple linear regression with AR(1) errors.

(a). Replace the AR(1) errors with an AR(2) process

References

1. Cowpertwait, P.S.P. and Metcalfe, A.V. (2009). *Introductory Time Series with R.* Springer, New York.
2. Cryer, J.D. and Chan, K.S. (2008). *Time Series Analysis: with Applications in R.* Springer, New York.
3. Brillinger, D.R. (2001). *Time Series: Data Analysis and Theory.* Society of Industrial and Applied Mathematics, Philadelphia, PA.

4. Box, G.E.P. and Jenkins, G.M. (1976). *Times Series Analysis Forecasting and Control.* Holden-Day, San Francisco, CA.

5. Bloomfield, P. (1976). *Fourier Analysis of Time Series, An Introduction.* John Wiley & Sons Inc., New York.

6

Time Series and Stationarity

As presented earlier, time series will often have a well-defined trend and seasonal components, and a good fitting model will account for these components in such a way that the residuals will tend to have means zero, constant variance, and of course autocorrelation. On the one hand, the adjacent observations have correlations that can be positive or negative, as for example, the airline passenger data where higher values are followed by higher values. On the other hand, adjacent observations can be negatively correlated such as, for example, when higher sales values are followed by lower numbers, etc. The models to be discussed will serve as the residuals with complex autocorrelation patterns for time series regression model. Refer to Broemeling[1] for an introduction to the Bayesian approach for the analysis of time series, and to Cowpertwait and Metcalfe,[2] who use R to generate realizations from various time series models.

6.1 Moving Average Models

The first class of model to be studied is the moving average, which is a special case of a stationary stochastic process. A stationary process $\{Y(t), t \geq 0\}$ is defined as one that satisfies the time invariance property:

$$[Y(t_1), Y(t_2), \ldots, Y(t_n)] \sim [Y(t_1 + h), Y(t_2 + h), \ldots, Y(t_n + h)] \qquad (6.1)$$

for all h>0.

There are processes that are stationary in the mean and covariance stationary, but not strictly stationary in the sense of (6.1). However, if a Gaussian process is covariance stationary, it is also strictly stationary.

A moving average process MA(q) is defined as

$$Y(t) = W(t) + \beta_1 W(t - 1) + \ldots + \beta_q W(t - q) \qquad (6.2)$$

where $W(t), W(t - 1), \ldots, W(t - q)$ is a sequence if independent white noise random variables with variance σ^2, and the $\beta_i, i = 1, 2, \ldots, q$ are unknown real parameters. It is obvious that the mean value function of the process is zero, the variance is

$$Var[Y(t)] = \sigma^2 (1 + \sum_{i=1}^{i=q} \beta_i^2).$$

(6.3)

and the autocorrelation of lag k is

$$\rho(k) = \sum_{i=0}^{i=q-k} \beta_i \beta_{i+k} / \sum_{i=0}^{i=q} \beta_i^2.$$

(6.4)

The posterior analysis for the MA(1) process is derived as follows. As an example, consider the MA(1) process

$$Y(t) = W(t) - \beta W(t-1), t = 1, 2, \ldots, n$$

(6.5)

Then, the process can be represented by

$$y = A(\beta)w$$

(6.6)

where the white noise vector with $n+1$ components is

$$w = (w_o, w_1, \ldots, w_n)$$

(6.7)

and

$w(t) \sim nid(0, \tau)$, where τ is the precision and $t = 0, 1, 2, \ldots, n$. Also $A(\beta)$ is the (n+1) x (n+1) matrix

$$A(\beta) = \begin{pmatrix} -\beta, 1, 0, 0, \ldots .0 \\ 0, -\beta, 1, 0, \ldots .0 \\ 0, 0, -\beta, 1, 0, .0 \\ \cdot \\ 0, 0, \ldots \ldots, -\beta, 0 \end{pmatrix}.$$

(6.8)

Thus, the n+1 vector

$$w \sim N[0, \tau I_{n+1}],$$

and it follows that y follows the multivariate normal

$$y \sim N\left\{0, \tau^{-1}[A(\beta)A^t(\beta)]^{-1}\right\}.$$

(6.9)

Also note that

$$A(\beta)A^t(\beta) = \begin{pmatrix} 1+\beta^2, -\beta, 0, \ldots \ldots \ldots, 0 \\ -\beta, 1+\beta^2, -\beta, 0, \ldots, 0 \\ \cdot \\ \cdot \\ \cdot \\ 0, 0, \ldots \ldots \ldots -\beta, 1+\beta^2 \end{pmatrix}, \qquad (6.10)$$

which is tri-diagonal with principal diagonal element $1+\beta^2$ and super- and sub-diagonal value $-\beta$.

Therefore, it can be shown that the joint density of the observations is

$$f(y|\beta,\tau) \propto \tau^{n/2}|A(\beta)A^t(\beta)|^{-1/2} \exp -(\tau/2)y^t[A(\beta)A^t(\beta)]^{-1}y. \qquad (6.11)$$

Now observe that $A(\beta)A^t(\beta)$ is positive definite and symmetric; thus, there exists a matrix Q, such that

$$D(\beta) = Q^t A(\beta)A^t(\beta)Q$$

is diagonal with i-th diagonal element

$$d_i(\beta) = 1 + \beta^2 - 2\beta \cos i\pi(n+1)^{-1} \qquad (6.12)$$

According to Gregory and Karney,[3]Q is independent of β and has ij-th element $\sqrt{2/(n+1)} \sin ij(n+1)^{-1}$.

Thus, it is seen that

$$A(\beta)A^t(\beta) = QD(\beta)Q^t \qquad (6.13)$$

and that the above density reduces to

$$f(y|\beta,\tau) \propto \tau^{n/2} \prod_{i=1}^{i=n} d_i(\beta)^{-1/2} \exp -(\tau/2)y^t QD^{-1}Q^t y, \ y \in R^n, \qquad (6.14)$$

which is a simplification over the representation (6.11) because the inverse of the precision matrix is in terms of its known eigenvalues and eigenvectors. What are the marginal posterior distributions of β and τ?

Of course, it is necessary to assign prior information to the two unknown parameters. To this end, assume that β and τ are independent a priori and that β has an arbitrary density $\zeta(\beta)$, but that the prior density for τ is the gamma density

$$\zeta(\tau) \propto \tau^{\alpha-1} \exp{-\tau\beta}, \ \tau > 0$$

where α and β are known hyperparameters.

It is obvious that the posterior density of the two parameter is

$$\zeta(\beta,\tau|y) \propto \zeta(\beta)\tau^{(n+2\alpha)/2-1} \prod_{i=1}^{i=n} d_i^{-1/2}(\beta)\exp{-(\tau/2)}$$

$$[2\beta + y^t QD^{-1}(\beta)Qy] \tag{6.15}$$

where $\beta \in R$ and $\tau > 0$.

Now integrating (6.15) with respect to τ gives the marginal posterior density of β as

$$\zeta(\beta|y) \propto \zeta(\beta) \prod_{i=1}^{i=n} d_i^{-1/2}(\beta) / [2\beta + y^t QD^{-1}(\beta)Qy]^{(n+2\alpha)/2}, \ \beta \in R. \tag{6.16}$$

Thus, based on (6.16), it is easy to plot the density of β and to find its moments numerically.

However, integration of (6.15) with respect to β is analytically intractable and can be achieved by numerical integration or resampling techniques.

Of course, since we know the variance–covariance matrix of the vector of observation, it is relatively easy to use WinBUGS for the Bayesian analysis.

As our first example for analyzing a MA(1) process, consider the process

$$Y(t) = W(t) + \beta_1 W(t-1), t = 2, 3, 4, \ldots\ldots, 100. \tag{6.17}$$

where $\beta_1 = .8$ and $\sigma^2 = 1$.

The following R code generates 1,000 observations from the MA(1) process (6.17).

RC 6.1

```
set.seed(1)
b<-c(.8)
x<-w<-rnorm(1,000)
for ( t in 2:1,000){
for ( j in 1:1) x[t]<-w[t]+b[j]*w[t-j]}
```

The first 20 values generated appear as components of the vector y.

Y=(−0.626453811, −0.317519724, −0.688713953, 0.926777912, 1.605732414, −0.556862167, −0.168945655, 1.128267947, 1.166441116, 0.155236694,

1.267470459, 1.599268171, −0.309365991, −2.711692352, −0.646828992,
0.855011125, −0.052137150, 0.930884000, 1.576290164, 1.250878277).

Based on these 20 values, the Bayesian analysis will estimate the parameters
of the MA(1) process (6.17), and the posterior analysis is executed with **BC 6.1**
with 35,000 observations for the simulation and a burn-in of 5,000. Noninformative prior distributions are used for the parameters: β_1 is normal(.8,.01),
while for σ^2 is gamma(.001,.001). Note it is assumed that the data vector of 20
observations has a multivariate normal distribution with mean vector zero
and a variance covariance matrix given by (6.3) and (6.4), respectively.

BC 6.1.

```
model;
{
beta~dnorm(.8,1)
v~dgamma(.1,.1)

for( t in 1:20){mu[t]<-0}
Y[1,1:20]~dmnorm(mu[],tau[,])

for( i in 1:20){Sigma[i,i]<-v*(1+pow(beta,2))}
for ( i in 1:19){Sigma[i,i+1]<-v*beta}
for ( i in 1:18){for ( j in i+2: 20) {Sigma[i,j]<-0}}
for ( i in 2:20){Sigma[i,i-1]<-v*beta}
  for( i in 3:20){ for ( j in 1:i-2 ){Sigma[i,j]<-0}}
tau[1:20,1:20]<-inverse(Sigma[,])
}
list(Y=structure(.Data=c(  −0.626453811,    −0.317519724,    −0.688713953,
0.926777912,   1.605732414,   −0.556862167,   −0.168945655,   1.128267947,
1.166441116,   0.155236694,   1.267470459,   1.599268171,   −0.309365991,
−2.711692352,   −0.646828992,   0.855011125,   −0.052137150,   0.930884000,
1.576290164, 1.250878277),.Dim=c(1,20)))
list(v=1, beta=.8)
```

The posterior analysis for the MA(1) process is reported in Table 6.1.

TABLE 6.1

Posterior Analysis for MA(1) Process

Parameter	Value	Mean	SD	Error	2½	Median	97½
β_1	.8	1.189	.5342	.01519	.3447	1.196	2.366
σ^2	1	.658	.3811	.00869	.1552	.592	1.571

Note that the posterior 95% credible intervals for the parameters include the values of the parameters used to generate the 20 observations used for the data and appear in the list statement of BC 6.1. Additional details of the posterior analysis for the MA(1) process is left as several exercises for the student.

Before leaving the MA model, the MA(1) process will serve as errors for a regression model.

6.2 Regression Models with Moving Average Errors

Consider the following model:

$$Y(t) = \gamma_0 + \gamma_1 t + \gamma_2 t^2 + Z(t), t = 1, 2, \ldots, n \qquad (6.18)$$

where the errors follow the MA(1) process

$$Z(t) = W(t) + \beta W(t - 1), \qquad (6.19)$$

with $\gamma_0 = 1, \gamma_1 = 2, \gamma_2 = 3, \beta = .8$, and the variance of white noise is $\sigma^2 = 1$.

The following R code generates 20 observations from the regression process (6.6) with the known parameters values as given above.

RC 6.2

```
set.seed(1)
b<-c(.8)
y<-x<-w<-rnorm(20)
for ( t in 2:20){
for ( j in 1:1) x[t]<-w[t]+b[j]*w[t-j]}
{for( t in 1:20)y[t]<-−1+2*t+3*t^2+x[t]}
```

Y=(5.373546, 16.682480, 33.311286, 57.926778, 87.605732, 120.443138, 161.831054, 210.128268, 263.166441, 321.155237, 387.267470, 458.599268, 533.690634, 614.288308, 705.353171, 801.855011, 901.947863, 1009.930884), 1123.576290, 1242.250878)

Of course, the goal of the Bayesian analysis is to estimate the unknown regression parameters $\gamma_i, i = 1, 2, 3$, the moving average coefficient β, and Gaussian noise variance σ^2. Using the above 20 values generated by RC 9.12 according to the model (6.6) and (6.7), **BC 6.2** is executed with 35,000 observations for the simulation and 5,000 for the burn-in. The 20 observations of vector Y are normally distributed with mean vector given by (6.6) and variance covariance matrix appropriate to the MA(1) time series errors parameter β.

BC 6.2.
model;{
beta~dnorm(.8,1)
v~dgamma(.1,.1)
g0~dnorm(1,1)
g1~dnorm(2,1)
g2~dnorm(3,1)
for(t in 1:20){mu[t]<-g0+g1*t+g2*t*t}
Y[1,1:20]~dmnorm(mu[],tau[,])
for(i in 1:20){Sigma[i,i]<-v*(1+pow(beta,2))}
for (i in 1:19){Sigma[i,i+1]<-v*beta}
for (i in 1:18){for (j in i+2: 20) {Sigma[i,j]<-0}}
for (i in 2:20){Sigma[i,i-1]<-v*beta}

for(i in 3:20){ for (j in 1:i-2){Sigma[i,j]<-0}} tau[1:20,1:20]<-inverse
(Sigma[,])
}

list(Y=structure(.Data=c(5.373546,   16.682480,   33.311286,   57.926778,
87.605732, 120.443138, 161.831054, 210.128268, 263.166441, 321.155237,
387.267470, 458.599268, 533.690634, 614.288308, 705.353171, 801.855011,
901.947863, 1009.930884, 1123.576290, 1242.250),.Dim=c(1,20)))
list(v=1, g0=1,g1=2,g2=3, beta=.8)

The posterior analysis is reported in Table 6.2.

Bayesian posterior means appear to be very close to the values of the parameters used to generate the data, which is most obvious for γ_2 with a posterior mean of 3.001 compared to its nominal value of 3!

As presented earlier, time series will often have a well-defined trend and seasonal components, and a good fitting model will account for these components in such a way that the residuals will tend to have means zero, constant

TABLE 6.2

Posterior Analysis for Regression Model with MA(1) Errors

Parameter	Value	Mean	SD	Error	2½	Median	97½
β	.8	1.184	.5182	.01273	.375	1.182	2.318
γ_0	1	.9514	.733	.02000	−.4733	.9477	2.393
γ_1	2	2.026	.1849	.00681	1.622	2.029	2.381
γ_2	3	3.001	.0093	.00033	2.982	3.001	3.018
σ^2	1	.7069	.411	.00804	.1643	.637	1.679

variance, and of course autocorrelation. On the one hand, the adjacent observations have correlations that can be positive or negative, as for example, the airline passenger data where higher values are followed by higher values. On the other hand, adjacent observations can be negatively correlated such as for example when higher sales values are followed by lower numbers etc. The models to be discussed will serve as the residuals with complex autocorrelation patterns for time series regression model.

Note that the posterior 95% credible intervals for the parameters include the values of the parameters used to generate the 20 observations used for the data and in the list statement of BC 6.2. Additional details of the posterior analysis for the MA(1) process is left as exercises for the student.

Bayesian posterior means appear to be very close to the values of the parameters used to generate the data, and this is most obvious for γ_2 with a posterior mean of 3.001 compared to its nominal value of 3!

6.3 Regression Model with MA Errors and Seasonal Effects

The next example is a regression model with MA(1) errors but with harmonic seasonal effects, which is specified as

$$Y(t) = \gamma_0 + \gamma_1 \sin(2\pi t/12) + \gamma_2 \sin(4\pi t/12) +$$
$$\gamma_3 \sin(8\pi t/12) + \gamma_4 \cos(8\pi t/12) + Z(t),$$
(6.20)

where

$$Z(t) = W(t) + \beta W(t-1)$$
(6.21)

is a MA(1) process with parameter β.

The Bayesian analysis will consist of estimating the parameters $\gamma_i, i=0,1,2,3,4$, β, and σ^2, the variance of the Gaussian noise process. The following R code generates 100 observations from the harmonic seasonal regression model (6.20) with parameter values $\beta = .8, \gamma_0 = 1 = \gamma_1, \gamma_2 = .2, \gamma_3 = \gamma_4 = .1$.

RC 6.3.

```
set.seed(1)
b<-c(.8)
y<-x<-w<-rnorm(100,0,.1)
for ( t in 2:100){
for ( j in 1:1) x[t]<-w[t]+b[j]*w[t-j]}
{for( t in 1:100)y[t]<-1+sin(2*pi*t/12)+0.2*sin(4*pi*t/12)+0.1*sin(8*pi*t/12)
+0.1*cos(8*pi*t/12)+x[t]}
```

The 100 values generated from the harmonic seasonal regression model are contained in the vectorY.

Y =(1.647162240, 1.870875972, 2.031128605, 1.822100655, 1.350765620,
1.044313783, 0.692913056, 0.283403931, 0.216644112, 0.012895725,
0.316939425, 1.259926817, 1.678871022, 1.631458709, 2.035317101,
1.814923976, 1.184978664, 1.193088400, 0.867436637, 0.295664964,
0.239409843, 0.149103876, 0.260219781, 0.907030029, 1.612642060,
1.946601130, 2.079930150, 1.569883984, 1.024717183, 1.103542152,
0.879110901, 0.268992728, 0.130544143, 0.023005281, 0.048182020,
0.948335779, 1.637179061, 1.865153408, 2.205257465, 1.893742468,
1.234794079, 1.061501944, 0.759235024, 0.282000526, 0.075657486,
−0.128477915, 0.170050963, 1.206019849, 1.760055634, 1.981751020,
2.210299206, 1.700068695, 1.175342237, 1.014353266, 0.762760944,
0.483259023, 0.221709844, −0.136419125, 0.163633572, 1.132072110,
1.939165029, 2.090833365, 2.165834736, 1.787402228, 1.118105231,
1.059417373, 0.544415142, 0.172735932, 0.232569723, 0.226893490,
0.411552265, 1.067046119, 1.714084542, 1.858076289, 1.899908849,
1.658276815, 1.169178890, 1.064647185, 0.717330182, 0.117572348,
−0.004028549, −0.061639304, 0.297186789, 1.041890280, 1.647316896,
1.983438676, 2.232946013, 1.784052458, 1.202859546, 1.156311384,
0.676923521, 0.247962315, 0.312669686, 0.160225630, 0.404892816,
1.282795319, 1.626827314, 1.743174026, 1.931677505, 1.584113791)

Based on the first 20 values generated from the regression model with harmonic seasonal effect, and using noninformative priors for the regression coefficients $\gamma_i, i = 0, 1, 2, 3, 4$, moving average parameter β, and variance of the Gaussian noise σ^2, the Bayesian analysis is executed with BC 6.3 using 35,000 observations for the simulation and 5,000 for the burn-in.

BC 6.3.

```
model;
{
g0~dnorm(0,.01)
g1~dnorm(0,.01)
g2~dnorm(0,.01)
g3~dnorm(0,.01)
g4~dnorm(0,.01)

  v~dgamma(.01,.01)
beta~dbeta(8,2)

for( t in 1:20){mu[t]<-g0+g1*sin(2*3.1416*t/12)+g2*sin(4*3.1416*t/12)+g3*sin
(8*3.1416*t/12)+g4*cos(8*3.1416*t/12)}
```

Y[1,1:20]~dmnorm(mu[],tau[,])

for(i in 1:20){Sigma[i,i]<-v*(1+pow(beta,2))}
for (i in 1:19){Sigma[i,i+1]<-v*beta}
for (i in 1:18){for (j in i+2: 20) {Sigma[i,j]<-0}}
for (i in 2:20){Sigma[i,i-1]<-v*beta}
    for(i in 3:20){ for (j in 1:i-2){Sigma[i,j]<-0}} tau[1:20,1:20]<-inverse
(Sigma[,])
}
list(Y=structure(.Data=c(1.647162240,   1.870875972,   2.031128605,
1.822100655,   1.350765620,   1.044313783,   0.692913056,   0.283403931,
0.216644112,   0.012895725,   0.316939425,   1.259926817,   1.678871022,
1.631458709,   2.035317101,   1.814923976,   1.184978664,   1.193088400,
0.867436637, 0.295664964),.Dim=c(1,20)))
list(g0=1,g1=1,g2=.2,g3=.1,g4=.1,beta=.8,v=.01))

Table 6.3 reports the results of the posterior analysis for the regression model is for the moving average errors.

Comparing the actual values of the parameters with their corresponding posterior means reveals the estimates are very close. In fact, the 95% credible intervals contain the actual values of the parameters used to generate the data in the list statement of BC 6.3, the sample information used for the Bayesian analysis.

6.4 Autoregressive Moving Average Models

One way to generalize the moving average and autoregressive processes is to combine the two into the ARMA(p,q) process defined as

$$Y(t) = \sum_{i=1}^{i=p} a_i Y(t-i) + W(t) + \sum_{j=1}^{j=q} \beta_j W(t-j) \qquad (6.22)$$

where $\{W(t), t>0\}$ is white noise with variance σ^2, $a_i, i = 1, 2, \ldots, p$ a sequence of unknown autoregressive parameters, and $\beta_j, j = 1, 2, \ldots, q$ as sequence of moving average parameters. As expected, the autocorrelation patterns are quite involved and represent complex correlation structure.

Consider an ARMA(1,1) process

$$Y(t) = \theta Y(t-1) + W(t) + \beta W(t-1), t = 2, 3, \ldots, n$$

TABLE 6.3

Posterior Analysis for Harmonic Seasonal Effects with MA(1) Errors

Parameter	Value	Mean	SD	Error	2½	Median	97½
β	.8	.7756	.1064	.00047	.5367	.7862	.9542
γ_0	1	1.046	.039	.00016	.9682	1.046	1.124
γ_1	1	.9134	.0533	.00021	.8077	.9132	1.019
γ_2	.2	.1414	.0475	.00018	.0475	.1413	.2363
γ_3	.1	.1043	.0273	.00010	.0500	.1043	.1589
γ_4	.1	.0946	.0284	.00011	.0382	.094	.1511
σ^2	.01	.0094	.0040	.00002	.0044	.0085	.0196

$$= W(t) + (\theta + \beta) \sum_{i=1}^{i=\infty} \theta^{i-1} W(t - i), \qquad (6.23)$$

and it follows that

$$Var[Y(t)] = \sigma^2 + \sigma^2(\theta + \beta)^2/(1 - \theta^2) \qquad (6.24)$$

and the autocovariance is

$$\text{cov}[Y(t), Y(t + k)] = (\theta + \beta)\theta^{k-1}\sigma^2 + (\theta + \beta)^2\sigma^2\theta^k/(1 - \theta^2). \qquad (6.25)$$

This is sufficient information for a Bayesian analysis in that the variance covariance matrix of the vector of observations Y can be coded in WinBUGS. The following R code generates observations from an ARMA(1,1) process with $\theta = .5, \beta = .5$, and $\sigma^2 = 1$.

RC 6.4

```
set.seed(1)
x<-arima.sim(n=10,000, list(ar=.5,ma=0.5))
coef(arima(x, order =c(1,0,1)))
```

The first 24 observations from the ARMA (1,1) process are the components of the vector Y:

Y=(2.00439112,   0.57587660,   −2.23738188,   −1.10110996,   −0.03302313,
−0.05516863, 0.90815676, 1.74721768, 1.87812076, 2.15498841, 2.31911919,
1.62519273,   −1.13947284,   −0.94458652,   −0.21850913,   −0.29311444,
−1.69520736, −2.06112993, −0.85169843, 1.14180112, 1.14745261, 0.91000405,
0.59503279, −1.10644568, −1.65674718)

Based on the above sample information and prior information for the unknown parameters β, θ, and σ^2, the Bayesian analysis is executed with 35,000 observations for the simulation and a burn-in of 5,000.

BC 6.4
```
model;
{
theta~dbeta(5,5)
beta~dbeta(5,5)
v~dgamma(.01,.01)
for ( t in 1:25){ mu[t]<-0}

Y[1,1:25]~dmnorm(mu[],tau[,])

for( i in 1:25){Sigma[i,i]<-v+v*pow(theta+beta,2)/(1-theta*theta)}

for( i in 1:24){for(j in i+1:25){Sigma[i,j]<-(theta+beta)*pow(theta,j-1)*v+
pow(theta+beta,2)*v*pow(theta,j)/(1-theta*theta)}}

for( i in 2:25){ for ( j in 1:i-1){Sigma[i,j]<-(theta+beta)*pow(theta,j-1)*v+
pow(theta+beta,2)*v*pow(theta,j)/(1-theta*theta)}}
    tau[1:25,1:25]<-inverse(Sigma[,])
}
list(Y=structure(.Data=c( 2.0039112, 0.57587660, −2.23738188, −1.10110996,
−0.03302313, −0.05516863, 0.90815676, 1.74721768, 1.87812076, 2.15498841,
2.31911919,    1.62519273,    −1.13947284,    −0.94458652,    −0.21850913,
−0.29311444,    −1.69520736,    −2.06112993,    −0.85169843,    1.14180112,
1.14745261,  0.91000405,  0.59503279,  −1.10644568,  −1.65674718          ),
.Dim=c(1,25)))

list(theta=.5,v=1,beta=.5)
```

Bayesian inferences for the parameters of the ARMA(1,1) model are reported in Table 6.4.

Comparing the posterior means to the actual value of the parameter reveals that the Bayesian analysis is providing sound inferences. For

TABLE 6.4

Posterior Distribution for ARMA(1,1)

Parameter	Value	Mean	SD	Error	2½	Median	97½
β	.5	.4979	.1512	.00071	.2094	.4978	.7857
θ	.5	.4604	.1407	.00081	.1956	.459	.7339
σ^2	1	1.01	.4248	.00386	.4092	.9369	2.05

example, consider β with actual value .5 and its posterior mean .4979, then one would conclude that indeed the Bayesian analysis is quite accurate, at least in this case.

6.5 Another Approach for the Bayesian Analysis of MA Processes

In the preceding approach to the analysis of the MA(1) process, the variance–covariance matrix of the generated observations was employed in the WinBUGS code. A different formulation will now be developed in order to perform the Bayesian analysis.

Consider the MA(1) model

$$Y(t) = W(t) - \phi W(t-1), t = 2, 3, \ldots \tag{6.26}$$

where the $W(t)$ are normal with mean 0 and unknown precision τ and moving average coefficient ϕ, where $|\phi| < 1$. The last restraint allows one to represent (6.13) as an infinite autoregressive process,

$$W(t) = Y(t) + \phi Y(t-1) + \phi^2 Y(t-2) + \phi^3 Y(t-3) + \ldots \tag{6.27}$$

Since $\phi| < 1$, the series converges and for a large power of ϕ, say m, the finite series

$$W(t) \sim Y(t) + \phi Y(t-1) + \phi^2 Y(t-2) + \phi^3 Y(t-3) + \ldots + \phi^m Y(t-m) \tag{6.28}$$

would be a good approximation to the MA(1) process (6.13).

Note that (6.14) can be expressed as

$$Y(t) = -(\phi Y(t-1) + \phi^2 Y(t-2) + \phi^3 Y(t-3) - \ldots) + W(t)$$

an autoregressive process of infinite order. Thus, for large m

$$Y(t) = -(\phi Y(t-1) + \phi^2 Y(t-2) + \phi^3 Y(t-3) - \ldots \phi^m Y(t-m)) + W(t) \tag{6.29}$$

would be a good approximation to the MA(1) process (6.26). Consider the following R code that generates 50 observations from a MA(1) process with $m=4$, $\phi = .5$, and $\tau = 1$.

RC 6.5
```
set.seed(1)
x<-w<-rnorm(50)
for ( t in 5:55){
  x[t]<-w[t]-.5*.5*w[t-1] -.5*.5*w[t-2]-.5*.5*.5*w[t-3]-.5*.5*.5*.5*w[t-4]}
```
The values appear below as vector y.

Y=(−0.626453811, −0.317519724, −0.688713953, 0.926777912, 1.605732414, −0.556862167, −0.168945655, 1.128267947, 1.166441116, 0.155236694, 1.267470459, 1.599268171, −0.309365991, −2.711692352, −0.646828992, 0.855011125, −0.052137150, 0.930884000, 1.576290164, 1.250878277)

The Bayesian analysis is executed with BC 6.5 using 70,000 observations for the simulation and 5,000 initial values.

BC 6.5
```
model;
{
theta~dbeta(5,5)
tau<-1/(1+pow(theta,2)+pow(theta,4)+pow(theta,6)+
pow(theta,8))
for( t in 1:20){ y[t]~dnorm(0,tau)}
}
list(y=c((   −0.626453811,   −0.317519724,   −0.688713953,   0.926777912,
1.605732414,   −0.556862167,   −0.168945655,   1.128267947,   1.166441116,
0.155236694,   1.267470459,   1.599268171,   −0.309365991,   −2.711692352,
−0.646828992,   0.855011125,   −0.052137150,   0.930884000,   1.576290164,
1.250878277))

list( theta=.5)
```

The posterior analysis is reported in Table 6.5.
The posterior mean of θ is .3639 compared to the true value of .5; however, the 95% credible interval is (.1609,.5737), which does include $\theta = .5$.

TABLE 6.5

Posterior Analysis for MA(1) Model

Parameter	Value	Mean	SD	Error	2½	Median	97½
θ	.5	.3639	.1072	.000424	.1609	.3623	.5737
τ	.75073	.8564	.0794	.000324	.6735	.8681	.9741

However, the posterior mean of τ is .8564 with a 95% credible interval of (.6735,.9741). Recall that this approach to the Bayesian analysis of the MA (1) model is only an approximation to the more reliable approach that assumes the 20 observations follow a multivariate normal distribution where the variance covariance matrix is reported in the BC 6.1. The student should compare Table 6.5 to Table 6.1 and examine how good the approximation is. The student will be asked to give a detailed explanation of the approximation.

6.6 Second-Order Moving Average Process

Consider the moving average process:

$$Y(t) = e(t) - \phi_1 e(t-1) - \phi_2\, e(t-2) \tag{6.30}$$

where the $e(t)$ are independent normal random variables with mean 0 and precision τ. It can be shown the variance is

$$Var[Y(t)] = \gamma_0 = (1 + \phi_1^2 + \phi_2^2)/\tau \tag{6.31}$$

and the lag one covariance is

$$Cov[Y(t), Y(t-1)] = \gamma_1 = (-\phi_1 + \phi_1\phi_2)/\tau \tag{6.32}$$

while the lag two covariance is

$$Cov[Y(t), Y(t-2)] = \gamma_2 = -\phi_2/\tau. \tag{6.33}$$

Thus, it is easy to show that the corresponding correlations are

$$\rho_1 = (-\phi_1 + \phi_1\phi_2)/(1 + \phi_1^2 + \phi_2^2) \tag{6.34}$$

and

$$\rho_2 = -\phi_2/(1 + \phi_1^2 + \phi_2^2). \tag{6.35}$$

Of course, $\rho_k = 0, k \geq 3$, and for the model

$$Y(t) = e(t) - e(t-1) + 0.6e(t-2)$$

it is easily shown that

$$\rho_1 = (-1 + (1)(-0.6)/(1 + 1^2 + (0.6)^2) = -1.6/2.36 = -0.678. \qquad (6.36)$$

As an example, consider the MA(2) process with

$\phi_1 = 1$, $\phi_2 = -0.6$, and $\tau = 1$. Download the TSA package in R and use the R command data(ma2.s). This will produce the following time series values:

Y=(−0.46745204,0.08148575,0.99380264, −2.69594173,2.81158699, −1.99705624, 0.37008349, −0.80087272,0.63257835,1.38196665, −1.51643286,1.23997582, −1.04837173,1.87956171, −2.15357656,1.04491984,0.92047573, −1.02093551, 1.15332582,0.85539738, −1.63234036, −0.03533793, −0.43116917,2.47092459, −3.24537059,2.18508568, −0.29724473, −0.40220659,1.34145825,1.28757378, 0.33948070,0.90135543, −0.14866024,1.21638447, −0.66271214, −1.04325070, 3.23533215, −2.73916295,2.16335010, −3.49338827,1.99719901,0.56919126, −0.71884815,1.16856201, −2.36115715,2.85693967, −0.83207487, −1.54954453, 1.11632535,1.70399944). I used the first 45 values.

Now use the R command plot(ma2.s) to plot 120 of the original time series values: See Figure 6.1.

These values will be used for the Bayesian analysis of the MA(2) process generated by RC 6.6, and the posterior analysis for this model is generated via BC 6.6 with 5,000 initial observations and 35,000 for the MCMC simulation.

BC 6.6

```
model;
{
Phi1~dnorm(1,1)
Phi2~dnorm(−.6,1)
v~dgamma(1,1)

for( t in 1:45){mu[t]<−0}
Y[1,1:45]~dmnorm(mu[],tau[,])

for( i in 1:45){Sigma[i,i]<-v*(1+pow(Phi1,2)+pow(Phi2,2))}
for ( i in 1:44){Sigma[i,i+1]<-v*(-Phi1+Phi1*Phi2)}
for( i in 1:43){Sigma[i,i+2]<- v*(-Phi2)}
for ( i in 1:41){for ( j in i+3: 45) {Sigma[i,j]<−0}}
Sigma[42,45]<−0
for ( i in 2:45){Sigma[i,i-1]<- v*(-Phi1+Phi1*Phi2)}
for(i in 3:45){Sigma[i,i-2]<-v*(-Phi2)}
  for( i in 4:45){ for ( j in 1:i-3 ){Sigma[i,j]<−0}}

  tau[1:45,1:45]<-inverse(Sigma[,])
```

```
}
list(Y=structure(.Data=c( −0.46745204, 0.08148575, 0.99380264, −2.69594173,
2.81158699, −1.99705624, 0.37008349, −0.80087272, 0.63257835, 1.38196665,
−1.51643286, 1.23997582, −1.04837173, 1.87956171, −2.15357656, 1.04491984,
0.92047573, −1.02093551, 1.15332582, 0.85539738, −1.63234036, −0.03533793,
−0.43116917, 2.47092459, −3.24537059, 2.18508568, −0.29724473, −0.40220659,
1.34145825, 1.28757378, 0.33948070, 0.90135543, −0.14866024, 1.21638447,
−0.66271214, −1.04325070, 3.23533215, −2.73916295, 2.16335010, −3.49338827,
1.99719901, 0.56919126,
−0.71884815, 1.16856201, −2.36115715 ),.Dim=c(1,45)))

list(v=1, Phi1= 1,Phi2=−.6)
```

The posterior analysis is reported in Table 6.6.

It is seen that the posterior median for ϕ_2 is −.678 with a 95% credible interval of (−2.668,−.2175) implying a good estimate of the lag 2 parameter. In a similar vein, the posterior analysis shows that the posterior median is a 'good' estimate of the corresponding parameter.

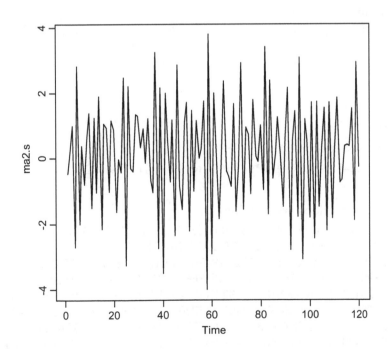

FIGURE 6.1
MA(2) Time Series with $\phi_1 = -1$, $\phi_2 = -0.6$, and $\tau = 1$

TABLE 6.6

Posterior Analysis for MA(2) Model

Parameter	Value	Mean	SD	Error	2½	Median	97½
V	1	.8672	.5404	.0363	.1766	.8586	1.892
ϕ_1	1	1.348	.5557	.0376	.5997	1.184	2.485
ϕ_2	−0.7	−1.169	.8146	.0552	−2.668	−0.678	−.2175

6.7 Quadratic Regression with MA(2) Residuals

Consider the model

$$Y(t) = \beta_0 + \beta_1 t + \beta_2 t^2 + e(t), \tag{6.37}$$

where

$$e(t) = W(t) - \phi_1 W(t-1) - \phi_2 W(t-2), \tag{6.38}$$

and the $W(t) \sim nid(0, \tau)$, $t = 1, 2, \ldots$

Also, the $\beta_i, i = 0, 1, 2$, $\phi_j, j = 1, 2$, and $\tau > 0$ are unknown parameters.

The following R code generates 45 observations from the MA(2) process (6.37). Assume the following values for the parameters: $\beta_0 = 1, \beta_1 = 2, \beta_2 = 3, \phi_1 = 1, \phi_2 = .6$ and $\tau = 1$.

RC 6.7.

```
>set.seed(1)
> .y<-x<-w<-rnorm(45)
>for( t in 3:45 ) {y[t]<-1+2*t+3*t^2+w[t]-w[t-1]-.6*w[t-2]}
>y
```

The 45 observations are as follows:

Y=(−0.6264538, 0.1836433, 33.3566003, 59.3207234, 85.2356041, 118.8928554, 163.1101928, 209.7431767, 261.5449992, 319.6758354, 387.4717007, 456.0612951, 532.0818475, 615.1726348, 709.7123752, 801.1589554, 901.3537848, 1009.9869866, 1121.8870991, 1240.2063784, 1365.8323433, 1496.5068181, 1632.7410423, 1774.4668015, 1928.5644385, 2081.5176565, 2241.5284378, 2407.7187204, 2583.0860796, 2762.7785430, 2947.2276280, 3135.2877678, 3333.6752516, 3536.6201960, 3744.4441425, 3961.9943480, 4182.8469403, 4409.5839733, 4643.3959127, 4880.6987384, 5124.4122854, 5376.4532565, 5635.0490392, 5897.0117168, 6164.3364031)

These observations will be used for the Bayesian analysis executed with BC 6.7. The analysis is executed with 45,000 for the simulation starting with 5,000. The prior distributions for the parameters are

BC 6.7.
```
model;
{
Phi1~dnorm(1,1)
Phi2~dnorm(-.6,1)
v~dgamma(1,1)
beta0~dnorm(1,.001)
beta1~dnorm(2,.001)
beta2~dnorm(3,.001)

for( t in 1:45){mu[t]<-beta0+beta1*t+beta2*t*t}
Y[1,1:45]~dmnorm(mu[],tau[,])

for( i in 1:45){Sigma[i,i]<-v*(1+pow(Phi1,2)+pow(Phi2,2))}
for ( i in 1:44){Sigma[i,i+1]<-v*(-Phi1+Phi1*Phi2)}
for( i in 1:43){Sigma[i,i+2]<- v*(-Phi2)}
for ( i in 1:41){for ( j in i+3: 45) {Sigma[i,j]<-0}}
Sigma[42,45]<-0
for ( i in 2:45){Sigma[i,i-1]<- v*(-Phi1+Phi1*Phi2)}
for(i in 3:45){Sigma[i,i-2]<-v*(-Phi2)}
  for( i in 4:45){ for ( j in 1:i-3 ){Sigma[i,j]<-0}}

  tau[1:45,1:45]<-inverse(Sigma[,])
}

list(Y=structure(.Data=c(-0.6264538,  0.1836433,  33.3566003,  59.3207234,
85.2356041,
118.8928554, 163.1101928, 209.7431767, 261.5449992, 319.6758354,
387.4717007, 456.0612951, 532.0818475, 615.1726348, 709.7123752,
801.1589554, 901.3537848, 1009.9869866, 1121.8870991, 1240.2063784,
1365.8323433, 1496.5068181, 1632.7410423, 1774.4668015, 1928.5644385,
2081.5176565, 2241.5284378, 2407.7187204, 2583.0860796, 2762.7785430,
2947.2276280, 3135.2877678, 3333.6752516, 3536.6201960, 3744.4441425,
3961.9943480, 4182.8469403, 4409.5839733, 4643.3959127, 4880.6987384,
5124.4122854,  5376.4532565,  5635.0490392,  5897.0117168,  6164.3364031),.
Dim=c(1,45)))

list(v=1,Phi1=1,Phi2=-.6, beta0=1, beta1=2, beta2=3)
```

Refer to Table 6.7 for the posterior analysis.

TABLE 6.7

Posterior Distributions for MA(2) Process

Parameter	Value	Mean	SD	Error	2½	Median	97½
ϕ_1	1	1.494	1.123	.03461	−1.64	1.645	3.179
ϕ_2	−.6	1.194	.4035	.01173	.3421	1.191	2.013
β_0	1	−2.697	1.183	.04032	−5.118	−2.651	−.4986
β_1	2	2.302	.1183	.00447	2.084	2.297	2.548
β_2	3	2.995	.00247	.000017	2.989	2.995	2.999
v	1	1.553	.8105	.02735	.5252	1.369	3.483

Based on Table 6.7, the student will be asked to provide statements that asses how well the posterior analysis does in estimating the parameters of the regression model with moving average errors.

6.8 Regression Model with MA(2) Errors and Seasonal Effects

The next example is a regression model with MA(2) errors but with harmonic seasonal effects, which is specified as

$$Y(t) = \gamma_0 + \gamma_1 \sin(2\pi t/12) + \gamma_2 \sin(4\pi t/12)$$
$$+ \gamma_3 \sin(8\pi t/12) + \gamma_4 \cos(8\pi t/12) + Z(t), \tag{6.39}$$

where

$$Z(t) = W(t) - \phi_1 W(t-1) - \phi_2 W(t-2), \tag{6.40}$$

is a MA(2) process with parameters $\phi_1 \phi_2$, and τ, where τ is the precision of the $W(t)$.

The Bayesian analysis will consist of estimating the parameters $\gamma_i, i = 0, 1, 2, 3, 4, \phi_1 \phi_2$, and τ, the precision of the Gaussian noise process. The following R code generates 100 observations from the harmonic seasonal regression model (6.39) with parameter values $\gamma_0 = 1 = \gamma_1, \gamma_2 = .2, \gamma_3 = \gamma_4 = .1,$
$\phi_1 = 1$, $\phi_2 = .6$, and $\tau = 1$.

RC 6.8.

```
set.seed(1)
y<-x<-w<-rnorm(45)
for ( t in 3:45){
x[t]<-w[t]-1*w[t-1]-.6*w[t-2]}
```

```
for( t in 1:45){y[t]<-1+sin(2*pi*t/12)+0.2*sin(4*pi*t/12)+0.1*sin(8*pi*t/12)
+0.1*cos(8*pi*t/12)+x[t]}
```

The 45 values are elements of the vector y.

Y=(1.08335381, 2.08627127, 1.45660035, 4.05014628, 0.42579652, −1.00714464,
1.82000039, 0.91375382, −0.35500078, −1.32679251, 1.66189312, 0.16129510,
−0.20834490, 0.07526270, 5.81237515, 1.88837827, 0.54397717, 2.08698664,
0.59690676, −0.62304446, −0.06765667, −0.49580981, −1.06876536, −1.43319846,
4.27424607, 2.42028447, 1.62843778, 0.44814323, 2.27627201, 2.87854305,
1.93743565, −1.54165508, −0.22474839, −0.38243196, −1.36566510, 2.09434802,
2.55674796, 2.48660124, 3.49591274, 1.42816128, −0.39752219, 0.55325647,
1.75884683, 0.18229397, −1.56359692)

These 45 values are plotted below in Figure 6.2.

These values are used in the list statement for the Bayesian analysis executed with BC 6.8. The number of MCMC simulated observations is 45,000 with 5,000 initial values.

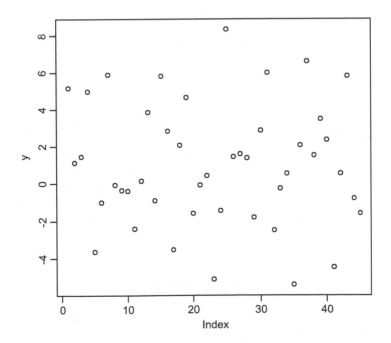

FIGURE 6.2
Simulated Values for Harmonic Regression

BC 6.8.
model;
{
Phi1~dnorm(1,1)
Phi2~dnorm(.6,1)
v~dgamma(1,1)
g0~dnorm(0,.01)
g1~dnorm(0,.01)
g2~dnorm(0,.01)
g3~dnorm(0,.01)
g4~dnorm(0,.01)

for(t in 1:45){mu[t]<-g0+g1*sin(2*3.1416*t/12)+g2*sin(4*3.1416*t/12)+g3*
sin(8*3.1416*t/12)+g4*cos(8*3.1416*t/12)}

Y[1,1:45]~dmnorm(mu[],tau[,])

for(i in 1:45){Sigma[i,i]<-v*(1+pow(Phi1,2)+pow(Phi2,2))}
for (i in 1:44){Sigma[i,i+1]<-v*(-Phi1+Phi1*Phi2)}
for(i in 1:43){Sigma[i,i+2]<- v*(-Phi2)}
for (i in 1:41){for (j in i+3: 45) {Sigma[i,j]<-0}}
Sigma[42,45]<-0
for (i in 2:45){Sigma[i,i-1]<- v*(-Phi1+Phi1*Phi2)}
for(i in 3:45){Sigma[i,i-2]<-v*(-Phi2)}
  for(i in 4:45){ for (j in 1:i-3){Sigma[i,j]<-0}}

  tau[1:45,1:45]<-inverse(Sigma[,])
}

list(Y=structure(.Data=c(1.08335381, 2.08627127, 1.45660035, 4.05014628,
0.42579652, −1.00714464, 1.82000039, 0.91375382, −0.35500078, −1.32679251,
1.66189312, 0.16129510, −0.20834490, 0.07526270, 5.81237515, 1.88837827,
0.54397717, 2.08698664, 0.59690676, −0.62304446, −0.06765667, −0.49580981,
−1.06876536, −1.43319846, 4.27424607, 2.42028447, 1.62843778, 0.44814323,
2.27627201, 2.87854305, 1.93743565, −1.54165508, −0.22474839, −0.38243196,
−1.36566510, 2.09434802, 2.55674796, 2.48660124, 3.49591274, 1.42816128,
−0.39752219,   0.55325647,   1.75884683,   0.18229397,   −1.56359692),.
Dim=c(1,45)))

list(v=1,Phi1=1,Phi2=.6, g0=1, g1=1, g2=.2,g3=.1,g4=.1)

The posterior analysis for this regression model is reported in Table 6.8.

TABLE 6.8

Posterior Analysis for Harmonic Regression with MA(2) Residuals

Parameter	Value	Mean	SD	Error	2½	Median	97½
ϕ_1	1	.5624	.7393	.03008	−1.108	.6494	1.906
ϕ_2	.6	.7638	.5507	.02496	.1492	.5942	2.263
γ_0	1	.9255	.07618	.00031	.7561	.929	1.077
γ_1	1	1.374	.1932	.00067	.9937	1.373	1.76
γ_2	.2	.4346	.3018	.00103	−.1638	.4367	1.03
γ_3	.1	.3881	.3434	.00123	−.2948	.3889	1.067
γ_4	.1	.1303	.3426	.00125	−.5411	.1288	.8043
v	1	.8147	.4279	.01945	.2137	.7453	1.78

The 95% credible interval for each parameter contains the value of that parameter used to generate the data. The posterior median for ϕ_2 is very close to its 'true' value of .6, as is the posterior median of γ_4.

6.9 Forecasting with Moving Average Processes

The following development of the predictive density can be found in Broemeling.[1, pp. 191–193] How does one forecast future observations with the MA(1) process? Recall for the autoregressive process the Bayesian predictive density was derived, which will be adopted for the moving average process. Consider the MA(1) model (6.5). The joint density of a future observation $y(n+1)$ and y, given β and τ is from (6.11)

$$f[y(n+1), y|\beta, \tau] \propto \tau^{(n+1)/2} \prod_{i=1}^{i=n+1} d_i(\beta)^{-1/2} \exp -(\tau/2)(y^t, y(n+1))$$
$$Q^* D^{-1}(\beta) Q^{*t}(y^t, y(n+1))^t \tag{6.41}$$

where $y \in R^n, y(n+1) \in r, \beta \in R$ and $\tau > 0$.
Also, note that

$$Q^{*t} A^*(\beta) A^{*t}(\beta) Q^* = D^*(\beta) \tag{6.42}$$

where $A^*(\beta)$ is the $(n+1)$ by $(n+1)$ analog of $A(\beta)$ (6.8), while Q^* is the analog of the diagonal matrix Q and has characteristic roots $1 + \beta^2 - 2\beta \cos i\pi(n+2)^{-1}, i = 1, 2, \ldots, n+1$ of $A^*(\beta)A^{*t}(\beta)$ as diagonal elements. The ij-th element of Q^* is $\sqrt{2/(n+2)} \sin ij(n+2)^{-1}$,

$i, j = 1, 2, \ldots, n + 1$. Consider the exponent of (6.41) and let

$$Q^* D^{*-1}(\beta) Q^{*t} = E(\beta) \tag{6.43}$$

which is partitioned as

$$E(\beta) = \begin{pmatrix} E_{11}(\beta), E_{12}(\beta) \\ E_{21}(\beta), E_{22}(\beta) \end{pmatrix} \tag{6.44}$$

where $E_{11}(\beta)$ is $n \times n$, and $E_{22}(\beta)$ is a scalar. Now the exponent of (6.41) can be expressed as

$$y^t E_{11}(\beta) y - 2y^t E_{12}(\beta) y(n+1) + y^2(n+1) E_{22}(\beta).$$

Multiply the density (6.41) by the prior density of β (a constant) and τ (a gamma with parameters α and γ), and integrate the product with respect to τ, then it follows that

$$f[y(n+1)|y, \beta] = \{y^2(n+1) E_{22}(\beta) - 2y(n+1) y^t E_{12}$$
$$(\beta) + y^t E_{11}(\beta) y + 2\gamma\}^{-(n+1+2\alpha)/2} \tag{6.45}$$

where $y(n+1) \in R, y \in R^n$. This is the conditional density of a future observation given the past n observations and the parameter β.

Now complete the square in the exponent of (6.45) by letting

$$A(\beta) = E_{22}(\beta),$$

$$B(\beta) = y^t E_{12}(\beta),$$

and

$$C(\beta) = y^t E_{11}(\beta) y + 2\beta.$$

Thus, the conditional density (6.45) may be expressed as

$$f[y(n+1)|y, \beta] \propto \{[y(n+1) - A^{-1}(\beta) B(\beta)]^2 A(\beta) +$$
$$C(\beta) - B^t(\beta) A^{-1}(\beta) B(\beta)\}^{-(n+1+2\alpha)/2}, \tag{6.46}$$

which is now recognized as a t-distribution with $n + 2\alpha$, location

$$E[y(n+1)|y, \beta] = A^{-1}(\beta) B(\beta), \tag{6.47}$$

and precision
$$P[y(n+1)|y,\beta] = A(\beta)(n+2\alpha)[C(\beta) - B^l(\beta)A^{-1}(\beta)B(\beta)]^{-1}. \tag{6.48}$$

It should be noted that this is a conditional density and not the marginal density of $y(n+1)$. To find the marginal distribution of the future observation, the conditional density must be averaged with respect the marginal posterior distribution of β (6.16). An alternative but suboptimal way to make inferences is to choose a value of β, and use the corresponding conditional inference via (6.46), (6.47), and (6.48).

6.10 Another Example

This example is taken from Cryer and Chan[4, pp. 57–59] and the model is

$$Y(t) = W(t) - \beta W(t-1), t = 1, 2, \ldots, n, \tag{6.49}$$

where $\beta = 0.6$ and $\iota = 1$ with $n-30$. The following R code was also taken from Cryer and Chan[4, pp. 57–59] and the code is given with RC 6.9

RC 6.9

```
> y<-w<-rnorm(30)
> for ( t in 2:30) y[t]<- w[t]-0.6*w[t-1]
> y
```

The R command

```
> y
```

generates the first 30 values appearing as components of the vector Y.

Y=(0.149591981, −1.432286670, 1.358821965, 1.257980997, −1.540857300, −1.480249550, 1.987565816, 1.060604813, −0.439238413, −1.605784483, 0.832018584, −0.817418718, −0.578563195, 2.960123654, 0.004398205, 0.100977621, 0.260697777, −1.307295462, 0.963172575, 0.735509439, 0.032628030, 0.655913707, −0.930255244, 2.659732385, −0.515045877, 1.284366781, −0.448217042, −0.711178862, 0.721295745, 0.502955835).
The following sample statistics are given by R as:

```
> mean(y)
[1] 0.03965288,
> sd(y)
[1] 1.29027, and the lag one autocorrelation is
> acf(y)$acf[2]
[1] −0.4101837.
```

The following R command generates a plot of the 100 values generated from the MA(1) process. See Figure 6.3.

>plot(y)

Based on the observations generated via RC 6.9, the Bayesian analysis is executed with BC 9.6 using 45,000 observations for the simulation and 5,000 of the initial group.

BC 6.9

```
model;
{
beta~dnorm(0.6,1)
v~dgamma(2,2)

for( t in 1:30){mu[t]<-0}
Y[1,1:30]~dmnorm(mu[],tau[,])

for( i in 1:30){Sigma[i,i]<-v*(1+pow(beta,2))}
for ( i in 1:29){Sigma[i,i+1]<-v*beta}
for ( i in 1:28){for ( j in i+2: 30) {Sigma[i,j]<-0}}
for ( i in 2:30){Sigma[i,i-1]<-v*beta}
for( i in 3:30){ for ( j in 1:i-2 ){Sigma[i,j]<-0}}
tau[1:30,1:30]<-inverse(Sigma[,])

}
list(Y=structure(.Data=c( 0.149591981, −1.432286670, 1.358821965, 1.257980997,
−1.540857300,
−1.480249550, 1.987565816, 1.060604813, −0.439238413, −1.605784483,
0.832018584, −0.817418718, −0.578563195, 2.960123654, 0.004398205,
0.100977621, 0.260697777, −1.307295462, 0.963172575, 0.735509439,
0.032628030, 0.655913707, −0.930255244, 2.659732385, −0.515045877,
1.284366781, −0.448217042, −0.711178862, 0.721295745, 0.502955835),.
  Dim=c(1,30)))
list(v=1, beta=.6)
```

The results of the posterior analysis are reported in Table 6.9.

The Bayesian analysis show that the posterior distribution is quite close to the values used to generate the data. For example, the posterior median of −.475 is compared to the true value of −.6 and the 95% credible interval for β is (−.2447,−.0097), which does indeed include the value −0.6. The same is true for the estimation of the precision of MA (1) process.

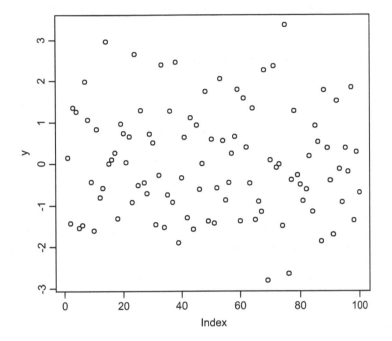

FIGURE 6.3
Cryer and Chan MA(1) Process

TABLE 6.9

MA(1) Posterior Analysis

Parameter	Value	Mean	SD	Error	2½	Median	97½
β	−.6	−.8231	.7536	.03593	−.2447	−.475	−.0097
v	1.073	.5034	.02221	1.134	2.037	1.134	2.037

6.11 Testing Hypotheses

As an example of testing hypotheses, consider the likelihood for the MA(1) process with likelihood

$$f(y|\beta,\tau) \propto \tau^{n/2}|A(\beta)A^t(\beta)|^{-1/2}\exp-(\tau/2)y^t[A(\beta)A^t(\beta)]^{-1}y. \tag{6.50}$$

Now consider the special case with density

$$f(y|\beta = 0,\tau) = \left[\tau^{(n+2)/2-1}/\Gamma((n+2)/2)\right]\exp(-\tau/2)\sum_{t=1}^{t=n}y_i^2, \tag{6.51}$$

which as a function of τ is recognized as a gamma distribution with parameter $((n+2)/2, \sum_{t=1}^{t=n} y_i^2/2)$.

Now consider a test of the null hypothesis

$$H : \tau = 1 \ versus the alternative \ A : \tau \neq 1. \tag{6.52}$$

A formal Bayesian test will be derived according to Lee.[5, ch. 4, pp. 126–129]

Recall that the special case of the MA(1) model

$$Y(t) = W(t) - \beta W(t-1), t = 1, 2, \ldots, n$$

with $\beta = 0$ is being considered, which is just a sequence of $nid(0, \tau)$ random variables; thus, we are testing the hypothesis that the sequence has precision one.

The following R program generates 50 observations from the process (6.5) with parameter vector $(\beta = 0, \tau = 1)$.

RC 6.10.

```
b<-0
y<-w<-rnorm(50)
for ( t in 2:50){y[t]<-w[t]-b*w[t-1]}
> y
y=(4959198, −1.34253148, 0.55330308, 1.58996284, −0.58687959, −1.83237731,
0.88813943, 1.348847, 0.51685467, −1.29567168, 0.05461558, −0.78464937,
−1.04935282, 2.33051196, 1.40270538, 0.94260085, 0.82625829, −0.81154049,
0.47624828, 1.02125841, 0.64538307, 1.04314355, −0.30436911, 2.47711092,
0.97122067, 1.86709918, 0.67204247, −0.30795338, 0.53652372, 0.82487007,
−0.96390148, −0.85508251, 1.88694694, −0.39181937, −0.98063295, 0.68733210,
−0.50504352, 2.15771982, −0.59979756, −0.69454669, 0.22392541, −1.15622333,
0.42241853, −1.32475526, 0.14108431, −0.53604800, −0.31160608, 1.55610964,
−0.44803329, 0.32112354).
```

It can be shown that $\sum_{t=1}^{t=50} y_i^2 = 74.79337$; thus, the posterior distribution of τ is gamma $(26,37.39665)$ with posterior mean .6952 and posterior variance .0185912.

According to Lee,[5, pp. 126–127] the posterior probability of the null hypothesis is

$$p_0 = 1/(1 + \lambda_1 p_1(y)/\lambda_0 p(y|\tau = 1)), \tag{6.53}$$

where λ_1 is the prior probability of the alternative hypothesis and the prior probability of the null is $\lambda_0 = 1 - \lambda_1$. Also,

$$p_1(y) = \int_0^\infty p_1(\tau)p(y|\lambda)d\tau, \qquad (6.54)$$

where

$$p_1(\tau) \propto 1/\tau, \tau > 0, \qquad (6.55)$$

is the prior density of τ under the alternative hypothesis.

It can be shown that the posterior probability of the null hypothesis (6.53) reduces to

$$p_0 = 1/[1 + [\lambda_1\Gamma(n/2)\exp(\sum_{i=1}^{i=n} y_i^2/2)/\lambda_0\Gamma((n+2)/2))$$

$$(\sum_{i=1}^{i=n} y_i^2/2)^{n/2}\ ]. \qquad (6.56)$$

Recall that $n=100$ and $\sum_{t=1}^{t=50} y_i^2 = 74.79337$; thus, there is sufficient information to calculate (6.56). The reader will be asked to compute (6.56) assuming $\lambda_0 = \lambda_1 = 1/2$.

6.12 Forecasting with a Moving Average Time Series

Assume the prior for β and τ is

$$\xi(\beta, \tau) \propto 1/\tau, \beta \in R, \tau > 0.$$

Recall from Section 6.2 that the posterior distribution of the moving average parameter β and τ is

$$\zeta(\beta|y) \propto \zeta(\beta) \prod_{i=1}^{i=n} d_i^{-1/2}(\beta)/[2\beta + y^t QD^{-1}(\beta)Qy]^{(n+2a)/2} \qquad (6.57)$$

and is nonstandard distribution and numerical integration is necessary to determine its characteristics. Based on Broemeling,[1, pp. 187–194] it can be shown that the conditional predictive density of a future observation given the sample

$s_n = [y(1), .., y(n)]$ and β is

$$f[y(n+1)|s_n, \beta] \propto [y^2(n+1)E_{22}(\beta) - 2\beta Y'E_{12}(\beta) + \\ Y'E_{11}(\beta)Y]^{-(n+1)/2}, \tag{6.58}$$

which can be simplified to

$$f[y(n+1)|s_n, \beta] \propto [y(n+1) - A^{-1}(\beta)B(\beta)]^2 + \\ C(\beta) - B'(\beta)A^{-1}(\beta)B(\beta)]^{-(n+1)/2}. \tag{6.59}$$

where $A(\beta) = E_{22}(\beta)$, $B(\beta) = Y'E_{12}(\beta)$, and $C(\beta) = Y'E_{11}(\beta)Y$, Furthermore,

$$E(\beta) = \begin{pmatrix} E_{11}(\beta), E_{12}(\beta) \\ E_{21}(\beta), E_{22}(\beta) \end{pmatrix} \tag{6.60}$$

and

$$E(\beta) = Q^*[D^*(\beta)]^{-1}Q^{*'} \tag{6.61}$$

and the ij-th element of Q^* is $\sqrt{2/(n+2)} \sin ij(n+2)^{-1}$. Note there exists a diagonal matrix Q such that $D(\beta) = Q'A(\beta)Q$ is diagonal with diagonal components.

$$d_i(\beta) = 1 + \beta^2 - 2\beta \cos i\pi(n+1)^{-1}, i = 1, 2, \ldots, n. \tag{6.62}$$

This is sufficient information to uniquely describe the conditional predictive density (6.59). We will not use this definition to compute the predictive density of $Y(n+1)$.

Consider the model

$$Y(t) = W(t) + \beta_1 W(t-1), t = 2, 3, 4, \ldots \ldots, 20. \tag{6.63}$$

where $\beta_1 = .8$ and $\sigma^2 = 1$, with RC 6.1. Let the Y vector contain the 20 observations, namely
Y=(−0.626453811, −0.317519724, −0.688713953,0.926777912,1.605732414,
−0.556862167, −0.168945655,1.128267947,1.166441116,0.155236694,1.267470459,
1.599268171,−0.309365991, −2.711692352, −0.646828992,0.855011125,
−0.052137150,0.930884000,1.576290164,1.250878277).
These observations are generated with RC 6.1.

To illustrate the Bayesian analysis for predicting $Y(n+1)$, note that

$$Y(n+1) = Y(21) = w(20) + \beta w(19),$$

where $w(i) \sim nid(0,1)$, i=1,2. Now consider the WinBUGS code

BC 6.10.
```
model;
{
beta~dnorm(.8,1)
v~dbeta(.5,1.5)

for( t in 1:20){mu[t]<-0}
Y[1,1:20]~dmnorm(mu[],tau[,])

for( i in 1:20){Sigma[i,i]<-v*(1+pow(beta,2))}
for ( i in 1:19){Sigma[i,i+1]<-v*beta}
for ( i in 1:18){for ( j in i+2: 20) {Sigma[i,j]<-0}}
for ( i in 2:20){Sigma[i,i-1]<-v*beta}
  for( i in 3:20){ for ( j in 1:i-2 ){Sigma[i,j]<-0}}
  tau[1:20,1:20]<-inverse(Sigma[,])
p21<-w[1]+beta*w[2]
w[1]~dnorm(0,v)
w[2]~dnorm(0,v)
}
list(Y=structure(.Data=c(   -0.626453811,   -0.317519724,   -0.688713953,
0.926777912,   1.605732414,   -0.556862167,   -0.168945655,   1.128267947,
1.166441116,   0.155236694,   1.267470459,   1.599268171,   -0.309365991,
-2.711692352,   -0.646828992,   0.855011125,   -0.052137150,   0.930884000,
1.576290164, 1.250878277),.Dim=c(1,20)))
list(v=1, beta=.8)
```

The Bayesian analysis is executed with 35,000 observations with 5,000 initial values and the results are reported in Table 6.10.

It is important to note that the future observation has a posterior distribution with median .0064 and 95% credible interval (-6.098, 6.127).

TABLE 6.10

Bayesian Prediction of Y[21]

Parameter	Value	Mean	SD	Error	2½	Median	97½
β	0.8	1.295	.5181	.01127	.4759	1.311	2.369
Y[21]	?	-.0010	3.006	.01538	-6.098	.006475	6.127
V	1	.5086	.213	.00327	.1581	.4914	.9201

The value of $\beta = 0.8$ used to generate the data for the simulation has a posterior median of 1.311 with 95% credible interval (.4759,2.369).

6.13 Exercises

1.   a. Define a stationary time series.

      b. Define a time series that is covariance stationary.

      c. If a time series is covariance stationary, is it also stationary?

      d. If a time series is stationary, is it also covariance stationary?

2.   a. Define a MA(1) process (6.5).

      b. What is the variance–covariance matrix of a MA(1) time series?

      c. Is the MA(1) time series stationary? Why? Explain your answer.

      d. What is the mean function of the MA(1) time series?

3. Based on Equation (6.11), derive the joint density (6.14) of n observations generated from the MA(1) series.

4. Based on the joint posterior density of (β, τ), derive the marginal posterior density of β (6.16).

5. Use RC 6.1 to generate 20 values from the MA(1) time series with $\beta = .8, \tau = 1$.

6.   a. Assume the prior distributions for the following parameters: $\beta \sim n(.8,.01)$ and $\tau \sim gamma(.001,.001)$, use BC 6.1 for the Bayesian analysis and execute the analysis with 35,000 for the simulation with 5,000 starting values. For data, use the 20 values generated by
RC 6.1.

      b. Compare your results for the posterior analysis with those reported in Table 6.1.

      c. What is the posterior mean and 95% credible interval for β?

7. Refer to (6.18) the quadratic regression model with MA(1) errors (6.19).

      a. Use RC 6.2 to generate 20 observations from the regression model (6.19) with the following parameter vector:

$$(\gamma_0 = 1, \gamma_1 = 2, \gamma_2 = 3, \tau = 1).$$

      b. Using the 20 values generated by RC 6.2 as data, and assuming the following prior distribution for the parameter values: $\gamma_0 \sim gamma(1,1), \gamma_1 \sim gamma(2,1), \gamma_2 \sim gamma(3,1), \tau \sim gamma$

(.1, .1) and $\beta \sim N(.8,1)$, execute a posterior analysis with 35,000 values for for the simulation and 5,000 starting values.

c. Compare your results with those reported in Table 6.2.

d. The value $\beta = .8$ was assumed for generating the 20 values with RC 6.2, how close is the posterior mean of β to this value?

8. Refer to the regression model (6.20) with seasonal effects and MA(1) residuals.

   a. Assuming $\gamma_0 = 1 = \gamma_1, \gamma_2 = 0.2, \gamma_3 = \gamma_4 = 0.1, \beta = 0.8$, and $\tau = 1$, use RC 6.3 to generate 100 observations from (6.20).

   b. Assume the following prior distributions for the parameters of the model: $\gamma_i \sim gamma(.01, .01), i = 0, 1, ..3, 4, \tau \sim gamma(.01, .01)$, and $\beta \sim beta(8,2)$, employ BC 6.3 to generate the posterior analysis with 35,000 for the simulation and 5,000 starting values.

   c. Compare your posterior results to that reported in Table 6.3.

   d. How close is the posterior median of β to the value $\beta = .8$ used to generate the data.

9.

   a. Define the ARMA(p,q) process (6.22).

   b. For the ARMA(1,1) model, derive the mean value function and the autocorrelation function.

   c. Assuming $\beta = .5, \theta = .5$, and $\tau = 1$, use RC 6.4 to generate 10,000 observations from the ARMA(1,1) series.

   d. Based on the first 24 observations generated via RC 6.4, use BC 6.4 for the posterior analysis. Assume the following prior distributions: $\theta \sim beta(5,5), \beta \sim beta(5,5)$, and $\tau \sim gamma(.01, .01)$.

   e. Execute the Bayesian analysis with BC 6.4, with 35,000 observations for the simulation and 5,000 starting values.

   f. Compare your posterior analysis to that reported in Table 6.4. Are your results similar? Explain your answer.

   g. What is the posterior mean of θ and how close is it to the value $\theta = .5$, the value used to generate the data generated by RC 6.4?

10. Refer to Section 6.6 and consider the MA(2) process defined by (6.30).

    a. Derive the variance function of the process.

    b. Derive the lag one correlation (6.34).

    c. Derive the lag two correlation (6.35).

11. Assume the MA(2) model has the parameter vector $(\phi_1 = 1, \phi_2 = -0.6, \tau = 1)$.

a. Using the R code (Ma2.s), generate 45 observations from the series (6.30). The R package TSA must be downloaded to execute this operation.

b. Generate a plot of the 45 values generated over time using the R command plot(y).

c. Assume the following prior distributions: $\phi_1 \sim N(1,1)$, $\phi_2 \sim N(-.6,1)$, and $\tau \sim gamma(1,1)$. Using the data generated above in part a, execute the Bayesian analysis with 45,000 observations for the simulation and 5,000 starting values.

d. Compare the results of your posterior analysis to that portrayed in Table 6.6.

e. The value of $\phi_1 = 1$ was used to generate the data; what is the posterior median and 95% credible interval for ϕ_1? How close is the posterior median to the value $\phi_1 = 1$?

12. Consider the quadratic regression (6.37) with MA(2) errors.

a. Assume the parameters have the following values: $\beta_0 = 1, \beta_1 = 2, \beta_2 = 3, \phi_1 = 1, \phi_2 = .6$, and $\tau = 1$. Use RC 6.7 and generate 45 observations from the series (6.37).

b. Using the 45 values generated by RC 6.7 as data, and assuming the following prior distributions: $\beta_0 \sim N(1,.001), \beta_1 \sim N(2,.001)$, $\beta_2 \sim N(3,.001), \phi_1 \sim N(1,1), \phi_2 \sim N(.6,1), \tau \sim gamma(1,1)$, execute the posterior analysis with 45,000 observations for the simulation with 5,000 starting values.

c. Compare your posterior analysis with that reported in Table 6.7.

d. What is the posterior mean for β_2? How close is the posterior mean to $\beta_2 = 3$, the value used to generate the data.

e. Does the 95% credible interval for β_2 include the value $\beta_2 = 3$?

13. Consider the regression model (6.39) with seasonal effects and MA (2) errors.

a. Assume the following values for the parameters: $\gamma_0 = 1 = \gamma_1$, $\gamma_2 = .2, \gamma_3 = \gamma_4 = 0.1, \phi_1 = 1, \phi_2 = .6$, and $\tau = 1$. Based on RC 6.8, generate 45 observation for the data.

b. With R command plot(y), plot the 45 values over time. See Figure 6.2.

c. Assume the following prior distributions for the parameters:-
$\gamma_0 \sim \gamma_1 \sim \gamma_2 \sim \gamma_3 \sim \gamma_4 \sim N(0,.01), \phi_1 \sim N(1,1), \phi_2 \sim N(.6,1)$, and $\tau \sim gamma(1,1)$.Now execute the Bayesian analysis with 45,000 for the simulation, and 5,000 starting values.

d. Compare your posterior analysis with that reported in Table 6.8. What is the 95% credible interval for γ_0?

Does it include 1?

14. Refer to Section 6.9, forecasting with the MA(1) series.

    a. Based on the joint conditional density (6.41), derive the predictive conditional expectation (6.47) for one future observation y [$n+1$].

    b. Derive the predictive conditional variance (6.48) for one future observation.

15. Consider the MA(1) model with $\beta = 0$, a test of the null hypothesis H $:\tau = 1$ versus A $:\tau \neq 1$ (6.51)

    a. Conduct a formal Bayesian test of H versus A using the description of Lee.[5, pp. 126–129]

    b. Use RC 6.10 to generate 50 observations from the MA(1) series with $\beta = 0$ and $\tau = 1$.

    c. Use these 50 observations and assume the prior probability of the null hypothesis is $\lambda_0 = 1/2$.

    d. Derive the posterior probability p_0 of the null hypothesis given by (6.55). Note that based on the 50 observations that $\sum_{t=1}^{t=50} y_i^2 = 74.79337$.

    e. Compute the actual value of p_0.

16. Refer to Section 6.11.

    (a) Use RC 6.1 to generate 20 observations from the MA(1) mode with β and τ.

    (b) Based on BC 6.10, execute the Bayesian analysis with 35,000 for the simulation and 5,000 initial values.

    (c) Compare you results with those reported in Table 6.10.

References

1. Broemeling, L.D. (1985). *Bayesian Analysis of Linear Models*. Marcel Dekker Inc., New York.
2. Cowpertwait, P.S.P. and Metcalfe, A.V. (2009). *Introductory Time Series with R*. Springer, New York.
3. Gregory, R.T. and Karney, D.I. (1969). *A Collection of Matrices for Testing Computational Algorithms*. Wiley Interscience, John Wiley & Sons Inc, New York.
4. Cryer, J.D. and Chan, K.S. (2008). *Time Series Analysis: with Applications in R*. Springer, New York.
5. Lee, P.M. (1997). *Bayesian Statistics, An Introduction, Second Edition*. John Wiley & Sons Inc., New York.

7

Time Series and Spectral Analysis

7.1 Introduction

This chapter introduces the reader to the Bayesian approach of estimating the spectral density of the basic times series model, including the autoregressive, the moving average, and the autoregressive moving averages models. The chapter begins with the fundamental ideas for defining the spectral density function. First, the general harmonic model containing sine and cosine functions is described for representing a seasonal time series. Next, the unit of measurement for frequency is defined which allows one to give a general definition of the spectral density. R plays an important role in generating time series and computing the spectral density. The spectrum function in R displays the original time series, computes the autocorrelation function, and finally computes the spectral density function. These ideas are illustrated with the elementary time series models that appear in Chapters 4 and 5. Using the WinBUGS programs appearing in Chapters 4 and 5, the Bayesian analysis of the spectral density function for the AR(1),AR(2),MA(1),MA(2), and ARMA (1,1) appears in Section 6.

7.2 The Fundamentals

From a historical point of view, spectral analysis was introduced as the search for hidden periodicities in time series data. We have seen in Chapter 4 the Bayesian analysis of regression models where the independent variables were sinusoidal harmonics of sine and cosine functions of time. The emphasis was on the time domain. In the case of spectral analysis, the emphasis will be on the frequency domain where the spectral function is a function of the frequency.

We review the fundamentals of working with trigonometric functions, and thus consider the cosine curve

$$R\cos(2\pi f t + \Phi) \tag{7.1}$$

where R is the amplitude, f the frequency, and Φ the phase. Note the curve repeats itself every $1/f$ time units and $1/f$ is called the period of the curve.

The following R code generates two cosine curves in discrete time from 1 to 96, and each has amplitude 1. The frequencies are 4/96 (a period of 4) and 14/96 (a period of .1458), respectively, where the lower frequency has phase 0 and the other has a phase of $.6\pi$.

RC 7.1

```
> win.graph(width=4.875,height=2.5,pointsize=8)
> t=1:96 # n=96
> cos1=cos(2*pi*t*4/96)
> cos2=cos(2*pi*(t*14/96+.3))
> plot(t,cos1, type='o', ylab='Cosines')
> lines(t,cos2,lty='dotted',type='o',pch=4)
```

The above plot command generates the two cosine curves exhibited in Figure 7.1.

See Cryer and Chan[1, ch. 13]

Now consider a time series with the two above cosine curves as the seasonal effects.

$$Y(t) = 2\cos[2\pi t 4/96] + 3\cos[2\pi(t14/96 + 0.3)] \tag{7.2}$$

where the first has amplitude 2 and the other amplitude 3. Now the periodicity is somewhat obscured, but spectral analysis provides the machinery for determining the hidden periodicities. The following R code generates 96 observations form (7.2)

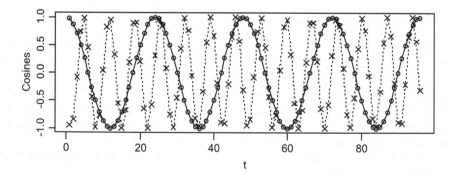

FIGURE 7.1
Two Cosine Curves

RC. 7.2.

```
> win.graph(width=4.875,height=2.5,pointsize=8)
> t=1:96 # n=96
> cos1=cos(2*pi*t*4/96)
> cos2=cos(2*pi*(t*14/96+.3))
> plot(t,cos1, type='o', ylab='Cosines')
> lines(t,cos2,lty='dotted',type='o',pch=4)
> # You may want to plot many of the graphs at fullscreen resolution to
    see more detail
> win.graph(width=4.875,height=2.5,pointsize=8)
> t=1:96 # n=96
> cos1=cos(2*pi*t*4/96)
> cos2=cos(2*pi*(t*14/96+.3))
> y=2*cos1+3*cos2
> plot(t,y,type='o',ylab=expression(y[t]))
```

The above plot command displays the 96 observations generated from (RC 7.2) as Figure 7.2.

Remember that the series (RC 7.2) does not have an error term, and of course if it did, the observed values would not follow the well-behaved sinusoidal curves.

Note that the series (7.20) is not in an ideal form for estimation of the parameters, thus consider the identity

$$R\cos(2\pi\,ft + \Phi) = A\cos(2\pi\,ft) + B\sin(2\pi\,ft), \qquad (7.3)$$

where
$$R^2 = A^2 + B^2, \qquad (7.4)$$

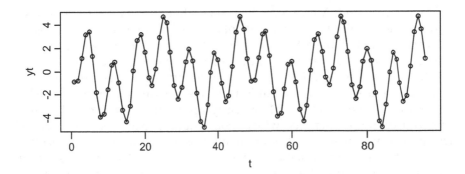

FIGURE 7.2
A Linear Combination of 2 Cosine Curves.

and

$$\Phi = \arctan(-B/A). \tag{7.5}$$

Therefore, for a fixed frequency f, $\cos(2\pi\, ft)$ and $\sin(2\pi\, ft)$ may be employed as predictor variables and the posterior distribution of A and B can be determined.

Now consider the time series of m cosine curves with arbitrary amplitudes and frequencies

$$Y(t) = A_0 + \sum_{j=1}^{j=m} A_j \cos(2\pi\, f_j t) + B_j \sin(2\pi\, f_j t) + w(t), t = 1, 2, \ldots, ntt \tag{7.6}$$

where the $w(t)$ are nid$(0,\tau)$.

How are the amplitudes to be estimated? Consider the special case where n is odd with $n=2k+1$, then the frequencies
$1/n, 2/n, \ldots, k/n$ are called Fourier frequencies, and the cosine and sine predictor variables at the frequencies are orthogonal and the least squares estimators are for $j=1,2,\ldots,m$,

$$\widehat{A}_j = (2/n) \sum_{t=1}^{t=n} Y(t) \cos(2\pi t j\,/\,n) \tag{7.7}$$

and

$$\widehat{B}_j = (2/n) \sum_{t=1}^{t=n} Y(t) \sin(2\pi t j\,/\,n). \tag{7.8}$$

What is the posterior distribution of the above parameters A_0, A_j, B_j, and τ? This is left as an exercise at the end of the chapter.

Cowperwait and Metcalfe,[2, ch. 9] state that because a time series is stationary and cannot have components at specific frequencies, it is possible to explain its behavior in terms of average frequency composition. Spectral analysis spreads the variance of the series over frequencies. For example, spectral analysis is especially an asset for describing wave-like phenomena which at first appear random but in fact have a frequency range over which the power is concentrated. Many early applications of spectral analysis involved economic activity which one would expect because economic activity always has seasonal components. Spectral analysis is very useful in the detection of period signals corrupted by noise. Astronomers use the spectrum to measure the red shift of galaxies and thus to estimate the relative velocity of the object. See Fiegelson and Babu[3, ch. 11] for the applications of spectral analysis to astronomical time and spatial series.

Any signal that has a repeating pattern is of course periodic with a period equal to the length of the pattern. Also, it is obvious that the sine wave plays the fundamental role in explaining the behavior of period signals. In fact, Fourier[4,5] demonstrated that linear combinations of sine waves provided adequate representation of most harmonic series and that spectral analysis is based on sine waves. It should be emphasized that the subject of spectral analysis can be confusing because the notation can differ between the various authors writing on the subject. For example, frequency can be given in radians or cycles per sampling interval. This can be clarified by defining frequency and cycle in terms of the unit circle as follows.

Imagine a circle with unit radius with center at (0,0) and suppose the radius rotates counterclockwise with rotation of ω radians per unit time; thus, let t be time, then the angle ωt in radians is the distance around the circumference from the positive real axis. If the radius completes a full circle, it has rotated 2π radians and the time (the period) taken for this one revolution is $2\pi/\omega$. Note that the sine function $\sin(\omega t)$ is the projection of the radius onto the vertical axis while the cosine function is the projection of the radius onto the horizontal axis at time t. Consider the sine function $A\sin(\omega t + \Phi)$ with frequency ω, amplitude A, and phase Φ, where the positive phase shift Φ denotes an advance of $\Phi/2\pi$ cycles, then it can be shown that

$$A\sin(\omega t + \Phi) = A\cos(\Phi)\sin(\omega t) + A\sin(\Phi)\cos(\omega t) \tag{7.9}$$

which is a fundamental result for the spectral analysis of a sampled sine wave of given amplitude and phase. Thus, the series can be easily fitted by linear regression with sine and cosine functions as independent variables.

7.3 Unit of Measurement of Frequency

The international measurement of frequency is hertz(Hz), which is one cycle per second and is equivalent to 2π radians per second. Note a frequency of f cycles per second is equivalent to ω radians per second, that is to say,

$$\omega = 2\pi f \Leftrightarrow f = \omega/2\pi \tag{7.10}$$

Now consider a time series of length $n\{Y(t), t = 1, 2, \ldots, n\}$, where n, is even; thus, consider fitting a regression model with $Y(t)$ as the response and $n{-}1$ predictor variables: $\cos(2\pi t/n), \sin(2\pi t/n), \cos(4\pi t/n), \sin(4\pi t/n), \ldots,$ $\cos(2(n/2 - 1)\pi t/n), \sin(2(n/2 - 1)\pi t/n), \cos(\pi t)$.

Now consider the model

$$
\begin{aligned}
Y(t) =&\, a_0 + a_1 \cos(2\pi t/n) + \beta_1 \sin(2\pi t/n) + a_2 \cos(4\pi t/n) \\
&+ \beta_2 \sin(4\pi t/n) + \ldots + a_{n/2-1} \cos(2(n/2 - 1)\pi t/n) \\
&+ \beta_{n/2-1} \sin(2(n/2 - 1)\pi t/n) + a_{n/2} \cos(\pi t).
\end{aligned}
\tag{7.11}
$$

Note that the number of observations is equal to the number of coefficients; thus, using linear regression to estimate the unknown coefficients
would leave 0 degrees of freedom for error and one would have a perfect
fit, where a_0 would be estimated by the overall mean $\bar{y}$. Also, note that the
lowest frequency depicted by the model is one cycle or 2π radians per
record length or equivalently $2\pi/n$ radians for sampling interval. The
reader is referred to Cowperwait and Metcalfe[2, pp. 173–175] for an excellent
description of this subject.

Now suppose with the model (7.11), that the r-th frequency is r cycles
per record length or $2\pi r/n$ radians per sampling interval, where
$1 \leq r \leq n/2$. Also, the largest frequency is π radians per sampling interval
of ½ cycles per sampling interval, which amounts to $n/2$ cycles in the
record length and alternating between +1 and –1 at each sampled point in
time. The model (7.11) is called the Fourier series for this time series. We
will refer to the sine wave that rotates r cycles in record length as the r-th
harmonic, and where the first harmonic as the fundamental frequency. The
following exposition is the key to understanding spectrum analysis.

The amplitude of the r-th harmonic is

$$
A_r = \sqrt{a_r^2 + \beta_r^2}
\tag{7.12}
$$

The fundamental result that is the foundation of spectral analysis is
Parseval's theorem, which represents the variance of the series as the sum
of $n/2$ components at integer frequencies varying from 1 to $n/2$ cycles per
record length.

In symbols, Parseval's theorem is given as

$$
\sum_{t=1}^{t=n} y^2(t)/n = A_0^2 + (1/2) \sum_{r=1}^{r=n/2-1} A_r^2 + A_{n/2}^2
\tag{7.13}
$$

or

$$
Var[Y(t)] = (1/2) \sum_{r=1}^{r=n/2-1} A_r^2 + A_{n/2}^2.
\tag{7.14}
$$

TABLE 7.1

Description of Period, Frequency, and Contribution to Variance by Harmonic

Harmonic	Period	Frequency cycles/ sampling interval	Frequency radians/ sampling interval	Contribution to variances
1	N	$1/n$	$2\pi/n$	$(1/2)A_1^2$
2	$n/2$	$2/n$	$4\pi/n$	$(1/2)A_2^2$
3	$n/3$	$3/n$	$6\pi/n$	$(1/2)A_3^2$
...	...			...
$n/2-1$	$n/(n/2-1)$	$(n/2-1)/n$	$(n-2)\pi/n$	$(1/2)A_{n/2-1}^2$
$n/2$	2	$1/n$	π	$A_{n/2}^2$

which follows from the fact that the sine and cosine terms in the time series regression (7.11) are uncorrelated and consequently that the variance of the sum of uncorrelated random variances is the sum of the individual variances. This is succinctly portrayed by Table 7.1.

7.4 The Spectrum

A plot of $A_r^2, r = 1, 2, ..$ versus r is the Fourier line spectrum, and the raw periodogram in R is obtained by joining the tips of the spikes in the Fourier line spectrum, which produces a continuous plot, where the scaling is done so that the area equals the variance. The periodogram distributes the variance over frequency, which according to Cowperwait and Metcalfe[2, pp. 175–177] has several disadvantages. The first is that the chosen frequencies are arbitrary, while the second is that the periodogram does not become smoother as the number of observations increases but includes more spikes packed loser together. To counteract this, the remedy is to smooth the periodogram by computing the moving average of spikes before joining the tips. The smoothed periodogram is also known as the sample spectrum, which is denoted as $C_{yy}()$, which had domain frequency given in radians ω or cycles per unit time. Another disadvantage is that smoothing will reduce the heights of the spikes; thus, additional smoothing blurs the features sought. Thus, one approach is present spectra with varying amounts of smoothing, which is easily implemented with the R function called *spectrum*.

The spectrum function in R has as an argument is the number of spikes in the moving average, which is a useful hint for the initial value for time series with a large number of observations.

The series should neither be mean adjusted before calculating the spectrum nor the coefficient a_0 of (7.11) should be set equal to 0 before averaging the spikes so as to avoid increasing the low-frequency contribution to the variance. In R, the function *spectrum* removes the linear trend before calculating the periodogram.

7.5 Examples

Our first example is to use R to calculate the spectrum of which noise with mean 0 and variance 1. The following R code RC 7.3 generates two sample spectra, the first is the raw periodogram, and the second the corresponding smoothed periodogram with a span of 65 for a smoother moving average. See Figure 7.3.

RC 7.3.

```
> layout(1:2)
> set.seed(1)
> y<-rnorm(2048)
> spectrum(y,log=c("no"))
```

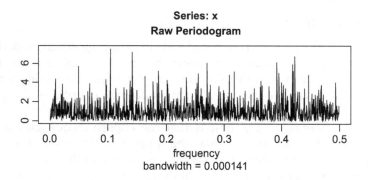

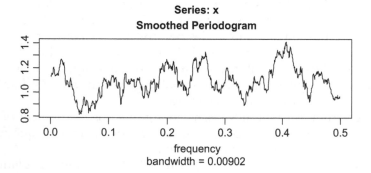

FIGURE 7.3
The Raw and Smoothed Periodogram of White Noise

> spectrum(y,span=65,log=c("no"))

The next example is that of an AR(1) series

$$Y(t) = \theta Y(t-1) + w(t), t = 2, 3, \ldots \tag{7.15}$$

with parameters $\theta = .9, \tau = 1$.
The following R code generates 1,024 observations from this process.

RC 7.4.
```
set.seed(1)
y<-w<-rnorm(1024)
for ( t in 2:1024){y[t]<-0.9*y[t-1]+w[t]}
layout(1:3)
plot(as.ts(y))
acf(y)
spectrum(y,span=65)
```

Figure 7.3 portrays the 1,024 values generated from the time series 7.15, the autocorrelation function, and the smoothed spectrum, with a span of 65 for a smoother moving average. See Figure 7.4.
The student will be asked to vary the span values in the spectrum function to see the effect on the smoothness.
Now consider the series

$$Y(t) = -0.9Y(t-1) + w(t), t = 2, 3, \ldots \tag{7.16}$$

which changes the autoregressive parameter θ from 0.9 to –0.9. What is the effect on the spectrum?
The R code below will generate 1,024 observations from the series (7.16).

RC 7.5
```
set.seed(1)
y<-w<-rnorm(1024)
for ( t in 2:1024){y[t]<-(-0.9)*y[t-1]+w[t]}
layout(1:3)
plot(as.ts(y))
acf(y)
spectrum(y,span=65)
```

Figure 7.5 reveals the dramatic shift in the spectrum
The above two examples of AR(1) will be revisited because one knows the parametric form of the spectrum. See Cryer and Chan.[1, ch. 13]

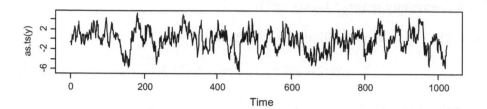

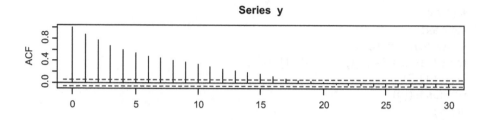

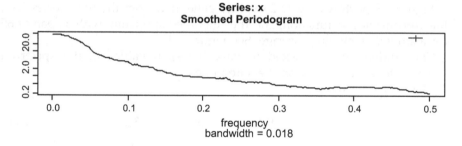

FIGURE 7.4
The Spectrum of the AR(1) Process

The two spectra should be compared, where the first is decreasing and the latter increasing over the frequency range.

The next example is for an AR(2) time series

$$Y(t) = Y(t-1) - 0.6Y(t-2) + w(t), t = 3, 4, \ldots \tag{7.17}$$

where the $\{w(t), t = 1, 2, \ldots\}$ are nid(0,1),

Consider the following R code for generating 1,024 values from the AR (2) series (7.17).

RC 7.6.
```
set.seed(1)
y<-w<-rnorm(1024)
```

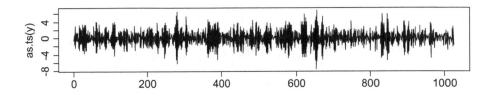

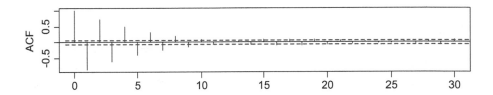

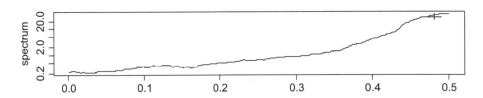

FIGURE 7.5
Spectrum of AR(1) Series with $\theta = -0.9$.

```
for ( t in 3:1024){ y[t]<-y[t-1]-0.6*y[t-2]+w[t]}
layout(1:3)
plot(as.ts(y))
acf(y)
spectrum (y, span=51)
```

The following figure reports the time series, autocorrelation function, and spectrum of (7.17).

7.6 Bayesian Spectral Analysis of Autoregressive Moving Average Series

Formulas are available for the spectrum of autoregressive processes, and this section is based on Chapter 13 of Cryer and Chan.[1]

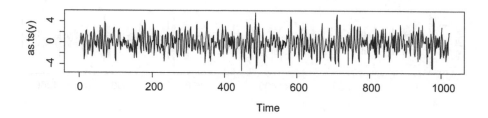

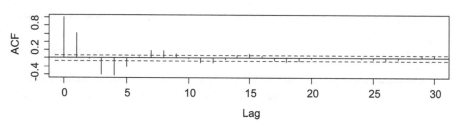

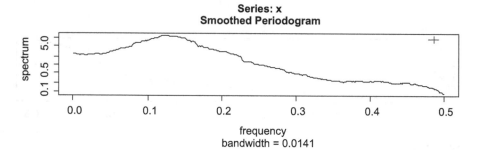

FIGURE 7.6
Spectrum of AR(2) Series

7.6.1 MA(1) Process

Consider the moving average series

$$Y(t) = W(t) + \beta_1 W(t-1), t = 2, 3, 4, \ldots\ldots, 100. \tag{7.18}$$

where $\beta_1 = .8$ and $\sigma^2 = 1$.

The following R code generates 1,000 observations from the MA(1) process (7.18).

RC 7.7
set.seed(1)

b<-c(.8)
x<-w<-rnorm(1000)
for (t in 2:1000){
for (j in 1:1) x[t]<-w[t]+b[j]*w[t-j]}
The first 20 values generated appear as components of the vector y.
Y=(−0.626453811, −0.317519724, −0.688713953, 0.926777912, 1.605732414, −0.556862167, −0.168945655, 1.128267947, 1.166441116, 0.155236694, 1.267470459, 1.599268171, −0.309365991, −2.711692352, −0.646828992, 0.855011125, −0.052137150, 0.930884000, 1.576290164, 1.250878277).

Based on these 20 values, the Bayesian analysis will estimate the parameters of the MA(1) process (7.18), and the posterior analysis is executed with **BC 7.1** with 35,000 observations for the simulation and a burn-in of 5,000. Noninformative prior distributions are used for the parameters: β_1 is normal(.8,.01), while for σ^2 is gamma(.001,.001). Recall that the data vector of 20 observations has a multivariate normal distribution with mean vector zero and a variance covariance matrix given by (6.4).

BC 7.1.

```
model;
{
beta~dnorm(.8,1)
v~dgamma(.1,.1)
for( t in 1:20){mu[t]<-0}
Y[1,1:20]~dmnorm(mu[],tau[,])
for( i in 1:20){Sigma[i,i]<-v*(1+pow(beta,2))}
for ( i in 1:19){Sigma[i,i+1]<-v*beta}
for ( i in 1:18){for ( j in i+2: 20) {Sigma[i,j]<-0}}
for ( i in 2:20){Sigma[i,i-1]<-v*beta}
  for( i in 3:20){ for ( j in 1:i-2 ){Sigma[i,j]<-0}}
  tau[1:20,1:20]<-inverse(Sigma[,])
pre<-1/v
f<-.5
# the spectral density function at f=1
sf<-(1+beta*beta-2*beta*cos(2*2.314*f))*v
}
list(Y=structure(.Data=c(   −0.626453811,   −0.317519724,   −0.688713953, 0.926777912,   1.605732414,   −0.556862167,   −0.168945655,   1.128267947, 1.166441116,   0.155236694,   1.267470459,   1.599268171,   −0.309365991, −2.711692352,   −0.646828992,   0.855011125,   −0.052137150,   0.930884000, 1.576290164, 1.250878277),.Dim=c(1,20)))
list(v=1, beta=.8)
```

It can be shown (see Cryer and Chan,[1, p. 333]) that the spectral density function for the MA(1) process is given by

$$s(f) = [1 + \beta^2 - 2\beta \cos(2\pi f)]/\tau \qquad (7.19)$$

where f is the frequency given in cycles per unit time and $0 \le f \le 1$, where $f=1$ corresponds to 2π radians.

The Bayesian analysis is executed via BC 7.1 using 70,000 observations for the simulation and 5,000 initial observations and the results are reported in Table 7.2. The goal is to estimate the spectral density function of the MA(1) process (7.19) at frequency $f=1$.

Note that the 'true' value of the spectral density at $f=1$ is 1.486 which is estimated as 2.089 with a 95% credible interval of (1.112,4.379).

Consider the above model from a different perspective by generating observations from (7.18) with the following R code taken from Cryer and Chan.[1, p. 333] You must upload the TSA package into the R program.

RC 7.8.

```
# MA1 spectral density
win.graph(width=4.875, height=2.5,pointsize=8)
beta=.8 # Reset theta for other MA(1) plots
ARMAspec(model=list(ma=-beta))
# R uses the plus convention in the MA specification.
```

The sample spectral density function appears in Figure 7.6.

The student will be asked to compute the spectral density (7.19) for $f=0,.1,.2,.3,.4,.5$, then using BC 7.1 to compare those results with the spectral density values reported in Figure 7.6. They should be very similar.

7.6.2 MA(2) Series

Consider the model

$$Y(t) = w(t) + \beta_1 w(t-1) + \beta_2 w(t-2), t = 3, 4, \dots \qquad (7.20)$$

TABLE 7.2

Posterior Analysis for the Spectral Density of MA(1)

Parameter	Value	Mean	SD	Error	2½	Median	97½
β	.8	1.193	.5335	.01084	.3574	1.198	2.364
S(f=1)	1.486	2.265	.8615	.0063	1.112	2.089	4.379
σ^2	1	.6561	.3802	.0062	.1556	.59	1.559

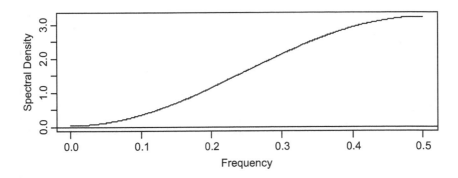

FIGURE 7.7
Spectral Density of MA(1)

where the $w(t)$ are independent normal random variables with mean 0 and precision τ.

It can be shown that the corresponding spectral density function at frequency f is

$$s(f) = [1 + \beta_1^2 + \beta_2^2 - 2\beta_1(1 - \beta_2)\cos(2\pi f) - 2\beta_2 \cos(4\pi f)]/\tau. \qquad (7.21)$$

Now consider the MA(2) series with $\beta_1 = 1$ and $\beta_2 = -0.6$, and the following R code that generates the spectrum corresponding to (7.21).

RC 7.9.
```
# MA(2) models
win.graph(width=4.875, height=2.5,pointsize=8)
beta1=1; beta2=-0.6 # Reset values of theta1 & theta2 for other MA(2)
models
ARMAspec(model=list(ma=-c(beta1,beta2)))
```

See Cryer and Chan[1, p. 334]for additional details.

The example of the MA(2) process with $\beta_1 = 1$, $\beta_2 = -0.6$, and $\tau = 1$ is continued. Download the TSA package in R and use the R command data (ma2.s). This will produce the following time series values

Y=(−0.46745204, 0.08148575, 0.99380264, −2.69594173, 2.81158699, −1.99705624, 0.37008349, −0.80087272, 0.63257835, 1.38196665, −1.51643286, 1.23997582, −1.04837173, 1.87956171, −2.15357656, 1.04491984, 0.92047573, −1.02093551, 1.15332582, 0.85539738, −1.63234036, −0.03533793, −0.43116917, 2.47092459,−3.24537059, 2.18508568, −0.29724473, −0.40220659, 1.34145825, 1.28757378, 0.33948070, 0.90135543, −0.14866024, 1.21638447, −0.66271214,

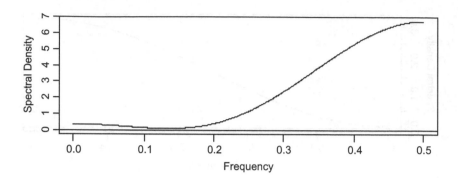

FIGURE 7.8
Sample Spectral Density Function for MA(2)

−1.04325070, 3.23533215, −2.73916295, 2.16335010, −3.49338827, 1.99719901, 0.56919126, −0.71884815, 1.16856201, −2.36115715, 2.85693967, −0.83207487, −1.54954453, 1.11632535, 1.70399944).

I used the first 45 values.

Now use the R command plot(ma2.s) to plot 120 of the original time series values:

These values will be used for the Bayesian analysis of the MA(2) process generated by RC 7.8, and the posterior analysis for this model is generated via BC 7.2 with 5,000 initial observations and 35,000 for the MCMC simulation.

BC 7.2.
```
model;
{
beta1~dnorm(1,1)
beta2~dnorm(-.6,1)
v~dgamma(1,1)
for( t in 1:45){mu[t]<-0}
Y[1,1:45]~dmnorm(mu[],tau[,])
for( i in 1:45){Sigma[i,i]<-v*(1+pow(beta1,2)+pow(beta2,2))}
for ( i in 1:44){Sigma[i,i+1]<-v*(-beta1+beta1*beta2)}
for( i in 1:43){Sigma[i,i+2]<- v*(-beta2)}
for ( i in 1:41){for ( j in i+3: 45) {Sigma[i,j]<-0}}
Sigma[42,45]<-0
for ( i in 2:45){Sigma[i,i-1]<- v*(-beta1+beta1*beta2)}
for(i in 3:45){Sigma[i,i-2]<-v*(-beta2)}
```

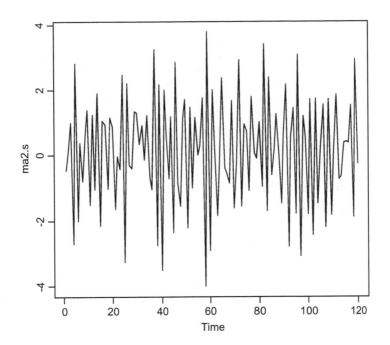

FIGURE 7.9
MA(2) Time Series with $\beta_1 = 1$, $\beta_2 = -0.6$, and $\tau = 1$

```
for( i in 4:45){ for ( j in 1:i-3 ){Sigma[i,j]<-0}}
tau[1:45,1:45]<-inverse(Sigma[,])
sf.3<-(1+pow(beta1,2)+pow(beta2,2)-2*beta1*(1-beta2)*cos(2*3.416*0.3)
-2*beta2*cos(4*3.416*0.3))*v
}
list(Y=structure(.Data=c( -0.46745204, 0.08148575, 0.99380264, -2.69594173,
2.81158699, -1.99705624, 0.37008349, -0.80087272, 0.63257835, 1.38196665,
-1.51643286, 1.23997582,
-1.04837173, 1.87956171, -2.15357656, 1.04491984, 0.92047573, -1.02093551,
1.15332582, 0.85539738, -1.63234036, -0.03533793, -0.43116917, 2.47092459,
-3.24537059, 2.18508568, -0.29724473, -0.40220659, 1.34145825, 1.28757378,
0.33948070, 0.90135543, -0.14866024, 1.21638447, -0.66271214, -1.04325070,
3.23533215, -2.73916295, 2.16335010, -3.49338827, 1.99719901, 0.56919126,
-0.71884815, 1.16856201, -2.36115715 ),.Dim=c(1,45)))
list(v=1, beta1= 1,beta2=-.6)
```

The Bayesian analysis for the MA(2) model is reported in Table 7.3.
One sees that the Bayesian analysis provides reasonable estimates for the parameters of the model.

TABLE 7.3

Posterior Analysis of MA(2) Series

Parameter	Value	Mean	SD	Error	2½	Median	97½
β_1	1	1.348	.5557	.0376	.5997	1.184	2.485
β_2	−0.6	−1.169	.8146	.0552	−2.668	−.678	−.2175
σ^2	1	.8672	.5404	.03635	.1766	.8586	1.892
$s(f=.3)$	1.56	3.714	1.107	.0282	2.123	3.53	6.39

7.6.3 The AR(1) Time Series

Consider the AR(1) model

$$Y(t) = \theta Y(t-1) + w(t), t = 2, 3, 4, \ldots \tag{7.22}$$

where $|\theta| < 1$ and $w(t) \sim n.i.d.(0, \tau), \tau < 0$.

According to Cryer and Chan,[1, p. 335] the spectral density function of this series at frequency f is
$s(f) = \sigma^2/[1 + \theta^2 - 2\theta\cos(2\pi f)]$, where

$$\sigma^2 = 1/\tau, \tau < 0, 0 > f > 0.5.$$

Based on the R code

Rc 7.10.

```
> # Exhibit 13.12 & 13.13
> # AR(1)
> win.graph(width=4.875, height=2.5,pointsize=8)
> phi=0.9 # Reset value of phi for other AR(1) models
> ARMAspec(model=list(ar=phi))
>
```

Cryer and Chan[1, p. 335] generate the following graph of the AR(1) process with parameters $\theta = 0.9, \tau = 1$ depicted in Figure 7.9, which is a decreasing function of frequency.

Note when $-1 > \theta > 0$, the spectral density is an increasing function of f.

The following R code RC 7.10 is given by Cowpertwait and Metcalfe[1, p. 81] which generates 100 values from the AR(1) process with autoregression coefficient $\theta = .6$. The autocorrelation function is acf while that for the partial autocorrelation is pacf.

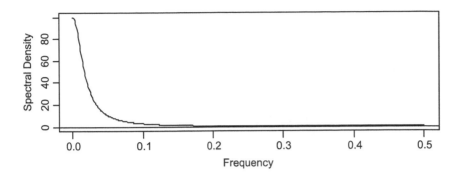

FIGURE 7.10
Spectral Density of AR(1) Series

RC 7.11.
```
set.seed(1)
y<-w<-rnorm(100)
for( t in 2:100) y[t]<-.6*y[t−1]+w[t]
time <- 1:100
plot(time,y)
acf(y)
pacf(y)
```

The first 50 values are labeled by the vector y= (−0.62645381, −0.19222896, −0.95096599, 1.02470121, 0.94432850,−0.25387129, 0.33510628, 0.93938847, 1.13941444, 0.37826027, 1.73873733 1.43308564, 0.23861080, −2.07153341, −0.11798913,−0.11572708, −0.08562651, 0.89246030, 1.35669738, 1.40791975, 1.76372922, 1.84037383, 1.17878928, −1.28207813, −0.14942113,−0.14578142, −0.24326436, −1.61671100, −1.44817665, −0.45096443, 1.08810089, 0.55007281, 0.71771530, 0.37682414, −1.15096507,−1.10557361, −1.05763412, −0.69389387, 0.68368905, 1.17338918, 0.53950991, 0.07034427, 0.73916994, 1.00016516, −0.08865660, −0.76068912, −0.09183151, 0.71343402, 0.31571420, 1.07053625)

Based on the representation (9.41) of an autoregressive process, the Bayesian analysis is repeated with **BC 7.3** executed with 35,000 observations for the simulation and a burn-in of 5,000. Note the prior for θ is beta (6,4) and that for the variance is expressed as a gamma(.001,.001) distribution. This induces a prior for variance covariance matrix of the 50 observations Y. The code is based on the fact that the 1 by 50 vector Y has a multivariate normal distribution with mean vector zero and covariance matrix given by (4.15) and (4.16).

BC 7.3
model;
{

v~dgamma(.01,.01)
theta~dbeta(6,4)

Y[1,1:50] ~ dmnorm(mu[], tau[,])
   for(i in 1:50){mu[i]<-0}

   tau[1:50,1:50]<-inverse(Sigma[,])

   for(i in 1:50){Sigma[i,i]<-v/(1-theta*theta)}
   for(i in 1:50){for(j in i+1:50){Sigma[i,j]<- v*pow(theta,j)*1/
   (1-theta*theta)}}
   for(i in 2:50){ for(j in 1: i-1){Sigma[i,j]<-v*pow(theta,i-1)*1/
   (1-theta*theta)}}

   sf.3<-v/(1+pow(theta,2)−2*theta*cos(2*3.416*0.3))

}

list(Y= structure(.Data=c(−0.62645381, −0.19222896, −0.95096599, 1.02470121,
0.94432850, −0.25387129, 0.33510628, 0.93938847, 1.13941444, 0.37826027,
1.73873733, 1.43308564, 0.23861080, −2.07153341, −0.11798913,−0.11572708,
−0.08562651, 0.89246030, 1.35669738, 1.40791975, 1.76372922, 1.84037383,
1.17878928, −1.28207813, −0.14942113,−0.14578142, −0.24326436, −1.61671100,
−1.44817665, −0.45096443, 1.08810089, 0.55007281, 0.71771530, 0.37682414,
−1.15096507,−1.10557361, −1.05763412, −0.69389387, 0.68368905, 1.17338918,
0.53950991, 0.07034427, 0.73916994, 1.00016516, −0.08865660, −0.76068912,
−0.09183151, 0.71343402, 0.31571420, 1.07053625),.Dim=c(1,50)))
list(theta=.6,v=1)

The result of the Bayesian analysis is reported in Table 7.4.

One sees the posterior analysis provides reasonable estimates of the unknown parameters of the AR(1) model. For example, the posterior median for the spectral density at f=.3 is .2895 compared to the 'true' value of .17 and the corresponding 95% credible interval is (.0901,.6986) which does indeed include the value .17. As for the autocorrelation parameter θ, the posterior estimate is quite close.

TABLE 7.4

Posterior Analysis for the AR(1) Series

Parameter	Value	Mean	SD	Error	2½	Median	97½
θ	.6	.626	.1509	.0021	.3101	.6385	.8682
σ^2	1	.5723	.2138	.0030	.2281	.5518	1.044
$S(f=.3)$	.17	.3104	.1615	.0022	.0901	.2805	.6986

7.6.4 Ar(2)

Consider the AR(2) process

$$Y(t) = \theta_1 Y(t-1) + \theta_2 Y(t-2) + w(t), t = 1, 2, \ldots \qquad (7.23)$$

where

$$w(t) \sim n.i.d.(0, \tau), t = 1, 2, \ldots$$

and

$$\sigma^2 = 1/\tau, \tau < 0.$$

The spectral density of an AR(2) time series (7.23) at frequency f is

$$s(f) = \sigma^2/[1 + \theta_1^2 + \theta_2^2 - 2\theta_1(1 - \theta_2)\cos(2\pi f) - 2\theta_2\cos(4\pi f)]. \qquad (7.24)$$

See Cryer and Chan[1, p. 336] for the derivation, where they also plot the spectral density for the AR(2) with parameters

$$\theta_1 = .31, \theta_2 = .32.$$

The following R code generates the spectral density of that autoregressive process, and the TSA package needs to be uploaded to the R worksheet.

RC 7.12

```
# AR(2)
win.graph(width=4.875, height=2.5,pointsize=8)
theta1=.31; theta2=.32 # Reset values of phi1 & phi2 for other AR(2) models
ARMAspec(model=list(ar=c(theta1,theta2)))
```

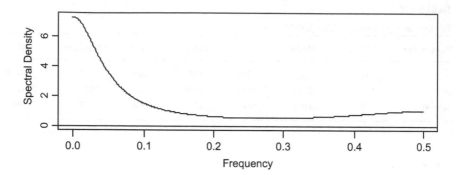

FIGURE 7.11
Spectral Density of AR(2) process.

See Figure 7.11 for a graph of the spectral density.

A Bayesian analysis is presented below using 50 observations for the data, which was generated from an AR(2) model with parameters $\theta_1 = .31, \theta_2 = .32$, and $\tau = 1$. Also, 135,000 observations were generated for the simulation with 5,000 starting values. Note that noninformative prior distributions were assigned to the parameters.

Observations

BC 7.4

```
Model;
{
  # AR(2) model
  mu[1] <- theta1*ym1+theta2*y0
  y[1] ~ dnorm(mu[1],tau)
  for(t in 2:50) {
  mu[t] <- theta1*y[t-1]+theta2*y[t-1]
  y[t] ~ dnorm(mu[t],tau)
  }
  # Prior distribution
  theta1 ~ dnorm(0.0, .001)
  theta2~ dnorm(0.01,.001)
  tau ~ dgamma(0.01, 0.01)
  y0 ~ dnorm(0.0,1.001)
  ym1 ~ dnorm(0.0,1.001)
sf.2<- 1/tau*(1+pow(theta1,2)+pow(theta2,2)−2*theta1*(1-theta2)*
cos(2*3.146*.2)−2*cos(4*3.146*.2))
```

}
# Simulated data: N=50, theta1=.31,theta2=.32, tau=1.00 and y0= 21.34232, ym1=22.
list(y=c(−0.62645381,   0.18364332,   −0.97916440,   1.35050570,   0.43483193,
−0.25350866, 0.54798759, 0.82707809, 1.00753159, 0.27161139, 1.91839081,
1.07146003, 0.32479709, −1.77114558, 0.67981086, −0.40095883, 0.07705197,
0.83941550,   1.10609663,   1.20540424,   1.64660361,   1.67831277,   1.12175510,
−1.10454753, 0.63637765, −0.21230688, −0.01796979, −1.54426122, −0.96262137,
−0.37463465, 0.93450397, 0.06702541, 0.70749076, 0.18696523, −1.09270329,
−0.69390371, −0.95906516, −0.57867278, 0.61373596, 0.76825861, 0.27003208,
0.07619102, 0.80699286, 0.83121211, −0.17284223, −0.49508837, 0.15579505,
0.65840111, 0.14161255, 1.13569597))
# Initial values
list(theta1=.31,theta2=.32, tau=1.0,y0=20.0,ym1=20)

The posterior analysis is reported in Table 7.5.

It can be seen that the 95% credible intervals include the 'true' value of the parameters, but it is obvious that the posterior distributions posses 'large' standard deviations.

7.6.5 ARMA(1,1) Time Series

Consider an ARMA(1,1) process

$$Y(t) = \theta Y(t-1) + W(t) + \beta W(t-1), t = 2, 3, \ldots, n \qquad (7.25)$$

and it follows that

$$Var[Y(t)] = \sigma^2 + \sigma^2(\theta + \beta)^2/(1 - \theta^2) \qquad (7.26)$$

and the autocovariance is

$$cov[Y(t), Y(t+k)] = (\theta + \beta)\theta^{k-1}\sigma^2 + (\theta + \beta)^2\sigma^2\theta^k/(1 - \theta^2). \qquad (7.27)$$

This is sufficient information for a Bayesian analysis in that the variance covariance matrix of the vector of observations Y can be coded in

TABLE 7.5

Posterior Analysis for AR(2)

Parameter	Value	Mean	SD	Error	2½	Median	97½
Sf(.2)	?	25.76	52.74	2.584	1.473	5.407	175.4
θ_1	.31	−.855	4.72	.2377	−12.59	−.0949	7.437
θ_2	.32	1.202	4.72	.2378	−7.072	.4407	12.95
τ	1	1.202	4.72	.2378	−7.072	.4407	12.95

WinBUGS. The following R code generates observations from an ARMA process (7.25).

It can be shown that the spectral density function at frequency f is

$$s(f) = [1 + \theta^2 - 2\theta\cos(2\pi f)]/\tau[1 + \beta^2 - 2\beta\cos(2\pi f)] \qquad (7.28)$$

The R code below generates observations from the ARMA(1,1) process with parameters $\theta = .5$, $\beta = .5$, and $\tau = 1$ and provides a plot of the spectral density function.

RC 7.13

```
> # ARMA(1,1)
> win.graph(width=4.875, height=2.5,pointsize=8)
> theta=0.5 ; beta=0.5 # Reset parameters for other ARMA(1,1) models
> ARMAspec(model=list(ar=beta,ma=-theta))
```
See Figure 7.12.

The following R code generates observations from an ARMA(1,1) process with, $\theta = .5$, $\beta = .5$, and $\sigma^2 = 1$.

RC 7.14.

```
set.seed(1)
x<-arima.sim(n=10000, list(ar=.5,ma=0.5))
coef(arima(x, order =c(1,0,1)))
```
The first 24 observations from the ARMA (1,1) process are the components of the vector Y:
Y=(2.00439112,   0.57587660,   −2.23738188,   −1.10110996,   −0.03302313, −0.05516863,   0.90815676,   1.74721768,   1.87812076,   2.15498841,   2.31911919, 1.62519273, −1.13947284, −0.94458652, −0.21850913, −0.29311444, −1.69520736,

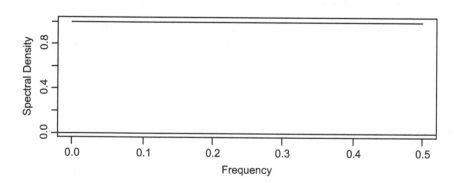

FIGURE 7.12
Spectral Density of an ARMA(1,1)

−2.06112993, −0.85169843, 1.14180112, 1.14745261, 0.91000405, 0.59503279, −1.10644568, −1.65674718)

Based on the above sample information and prior information for the unknown parameters β, θ, and σ^2, the Bayesian analysis is executed via BC 7.5 with 35,000 observations for the simulation and a burn-in of 5,000. Note the prior distributions for the parameters are beta(5,5) for the autoregressive and moving average parameters.

BC 7.5

```
model;
{
theta~dbeta(5,5)
beta~dbeta(5,5)
v~dgamma(.01,.01)
for ( t in 1:25){ mu[t]<-0}
Y[1,1:25]~dmnorm(mu[],tau[,])
for( i in 1:25){Sigma[i,i]<-v+v*pow(theta+beta,2)/(1-theta*theta)}
for( i in 1:24){for(j in i+1:25){Sigma[i,j]<-(theta+beta)*pow(theta,j-1)*v+
pow(theta+beta,2)*v*pow(theta,j)/(1-theta*theta)}}
for( i in 2:25){ for ( j in 1:i-1){Sigma[i,j]<-(theta+beta)*pow(theta,j-1)*v+
pow(theta+beta,2)*v*pow(theta,j)/(1-theta*theta)}}
   tau[1:25,1:25]<-inverse(Sigma[,])
sf.2<-(1+pow(theta,2)-2*cos(1.258))/v*(1+pow(beta,2)-2*cos(1.258))
}
list(Y=structure(.Data=c( 2.0039112, 0.57587660, −2.23738188, −1.10110996,
−0.03302313, −0.05516863, 0.90815676, 1.74721768, 1.87812076, 2.15498841,
2.31911919, 1.62519273, −1.13947284, −0.94458652, −0.21850913, −0.29311444,
−1.69520736, −2.06112993, −0.85169843, 1.14180112, 1.14745261, 0.91000405,
0.59503279, −1.10644568, −1.65674718 ),.
Dim=c(1,25)))
list(theta=.5,v=1,beta=.5)
```

The posterior analysis is reported in Table 7.6.

It is seen that the posterior analysis provides quite accurate estimates for the parameters, and indeed the 95% credible intervals include their 'true' values.

7.7 Sunspot Cycle

The periodic behavior of the sunspot cycle is evident from the figure below, and one can see the approximate period is 11 years. The first value corresponds to January 1749 and the last to May 2106. The images can be accessed at solarscience.msfc.nasa.gov/SunspotCycle.shtml.

TABLE 7.6

Posterior Analysis for the ARMA(1,1)

Parameter	Value	Mean	SD	Error	2½	Median	97½
β	.5	.4978	.1513	.0011	.2009	.4974	.7861
S(f=.3)	?	.5157	.396	.0042	.123	.4004	1.567
θ	.5	.4599	.1412	.0014	.1943	.4593	.7334
$\sigma^2 = 1/\tau$	1	1.012	.4261	.0043	.4139	.9365	2.046

The 'sunspotnumber' is then given by the sum of the number of individual sunspots and ten times the number of groups. Since most sunspot groups have, on average, about ten spots, this formula for counting sunspots gives reliable numbers even when the observing conditions are less than ideal and small spots are hard to see. . See Figure 7.13.

It will be left to the student to develop a statistical model for the sunspot cycle and to estimate the spectral density function by Bayesian techniques. See Edwards, Meyer, and Christiansen[6] for a Bayesian approach to estimating the spectrum of the sunspot cycle. An excellent account of spectrum analysis can be seen in Naidu[7].

7.8 Comments and Conclusions

This chapter introduces the Bayesian approach for the analysis of the spectrum of a time series. The posterior analysis of the spectrum for the basic time series such as the AR(1), AR(2), MA(1), and MA(2) series was explained. For such series, the spectral density is a function of the parameters of the model for that series, thus is easily estimated via the usual characteristics of its posterior distribution. Information about these models in the form of data was generated with R code and the posterior analysis implemented with WinBUGS. One interesting example is the sunspot cycle which was fitted with a sinusoidal model and the parameters of which were estimated by the appropriate posterior distribution. Of course, the subject matter can be expanded to include more complicated time series for periodic data.

7.9 Exercises

1. Write a three-page essay that summarizes Chapter 7.
2. Using RC 7.1, provide a plot of the two cosine curves portrayed in Figure 7.1.

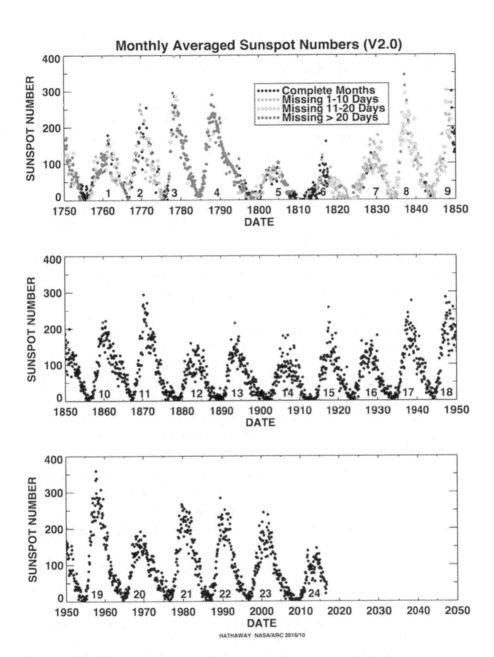

FIGURE 7.13
Sunspot Cycle

3. Based on the time series defined by (7.2), use RC 7.2 to plot the linear combination of two cosine curves portrayed in Figure 7.2.

4. What is the periodogram of a time series? How is it related to the spectral density function?

5. Refer to the AR(1) series (7.18) and employ RC7.4 to generate 1,024 observations where the parameters are $\theta = .9, \tau = 1$. Use the spectrum function in R to plot the spectrum reported in Figure 7.3. Remember to download the TSA package when using RC 7.4.

6. Refer to the AR(2) model defined by (7.17) and use RC 7.6 to generate 1,024 values from the series. Also, generate a plot of the autocorrelation function with the R command acf and the smoothed periodogram with the spectrum function.

7. (a) Based on the MA(1) model (7.18), use RC 7.7 to generate 1,000 observations from the series with parameters $\beta = .8, \tau = 1$.

   (b) Derive the spectral density function (7.19) for the MA(1) time series.

   (c) Use the first 20 observations generated by RC 7.7 as data for the Bayesian analysis executed with BC 7.1 with 35,000 observations for the simulation and 5,000 initial values.

   (d) Compare your results of the posterior analysis with that reported in Table 7.2. What are the posterior mean and 95% credible interval for the spectral density at frequency *f*=1?

8. Refer to the MA(1) series of part a of problem 7 above, and use RC 7.7 to generate a plot of the spectral density function portrayed by Figure 7.6.

9. Refer to RC 7.8.

   a. Generate the graph (Figure 7.7) of the spectral density function corresponding to the MA(1) series with parameters $\beta = .8, \tau = 1$.

   b. What is the value of the spectral density at *f*=.2?

10. Refer to the MA(2) time series model (7.20) with parameters $\beta_1 = 1, \beta_2 = -0.6, \tau = 1$.

   a. Derive the spectral density function given by (7.21).

   b. Use RC 7.9 to generate a graph of the spectral density function of Figure 7.8.

11. a. Use R to generate 45 observations from a MA(2) time series with parameters $\beta_1 = 1, \beta_2 = -0.6, \tau = 1$The R command data(ma2.s) must be used to generate the observations. Remember to download the TSA package in R.

   b. Use the 45 observations as data for the Bayesian analysis to be executed with BC 7.2 using 45,000 observations for the simulation and 5,000 initial values.

    c. Compare your posterior analysis to that reported in Table 7.3.

    d. What is the posterior median and 95% credible interval for β?

12. a. Consider the AR(1) time series (7.22) and derive the corresponding spectral density function.

    b. Assume the parameters of the series is $\theta = 0.9, \tau = 1$, and use RC 7.10 to generate a graph of the spectral density function given by Figure 7.10.

    c. Use the RC 7.11 to generate 100 observations from the AR(1) model with parameters $\theta = 0.6, \tau = 1$.

    d. Plot the 100 values and the corresponding acf.

13. Use the first 50 values generated in part d above for a Bayesian analysis.

    a. Use the 50 values as data and execute BC 7.3 with 45,000 for the simulation and 5,000 starting values.

    b. Compare your result to those posted in Table 7.4.

    c. What is the 95% credible interval for the spectral density at frequency *f*=0.3?

14. Refer to the AR(2) series (7.23).

    a. Derive the formula for the spectral density given by (7.24).

    b. With RC 7.12, generate the graph of the spectral density shown in Figure 7.11. Assume that the parameters of the process are $\theta_1 = .31, \theta_2 = .32, \tau = 1$ and remember to download the TSA package.

15. Referring to problem 14, perform a Bayesian analysis for the AR(2) model with parameters $\theta_1 = .31, \theta_2 = .32, \tau = 1$.

    a. Using R, generate 50 observations from the AR model.

    b. Using the 50 observations as data, execute the posterior analysis with BC 7.4.

    c. Are your results the same as those reported in Table 7.5?

    d. What is the 95% credible interval for the spectral density at frequency 0.2?

16. Refer to the ARMA(1,1) time series given by (7.25).

    a. Verify the spectral density given by (7.28).

    b. Assume the parameters of the model are $\theta = .5, \beta = .5, \tau = 1$ and with RC 7.14 generate 24 observations from the ARMA model.

    c. Use the 24 observations as data for the Bayesian analysis to be executed with BC 7.5, using 35,000 observations for the simulation with 5,000 initial values.

    d. Compare your results reported in Table 7.6.

    e. What is the posterior mean and standard deviation for β?

17. Refer to the sunspot cycle of Section 7.7.

    a. Download the sunspot cycle data consisting of monthly averages from: solarscience.msfc.nasa.gov/SunspotCycle.shtml.

    b. Construct a model for the sunspot cycle data making sure the model fits the data well.

    c. Based on your model, estimate the spectral density over (0,.5) using the spectrum function in R with TSA downloaded.

    d. Perform a Bayesian analysis with WinBUGS that estimates the parameters of your model.

References

1. Cryer, J.D. and Chan, K.S. (2008). *Time Series Analysis: with Applications in R.* Springer, New York.
2. Cowpertwait, P.S.P. and Metcalfe, A.V. (2009). *Introductory Time Series with R.* Springer, New York.
3. Fiegelson, E.D. and Babu, G.J. (2015). *Modern Statistical Methods for Astronomy with R Applications.* Cambridge University Press, London UK.
4. Fourier, J.B.J. (1826). "Memoire sur le resultats moyens deduits d'un grand nombre d'observations", *Recherches statistiques sur la ville de Paris et le department de la Seine*, 1826, Reprinted in Oeuvres, 2, 525–545.
5. Fourier, J.B.J. (1890). *Ouvres, G, Darboux Editor, 2 Volumes.* Gauthier-Villars, Paris.
6. Edwards, M.C., Meyer, R. and Christiansen, N. (2018)."Bayesian nonparametric spectral density estimation using B-splines priors." Online journal address: https://doi.org/10.1007/s1122-017-97969.
7. Naidu, P.S. (1995). *Modern Spectrum Analysis of Time Series.* CRC Press, Boca Raton, FL.

8

Dynamic Linear Models

8.1 Introduction

According to Broemeling,[1, ch. 6] linear dynamic models are used by communications and control engineers to monitor and control the state of a system as it evolves through time. In order to determine whether a system is operating satisfactory and later to control the system, one must know the behavior of the system at any time. For example, in the navigation of a submarine, spacecraft, or aircraft, the state of the system is the position and velocity of the craft. The state of the system may be random because physical systems are often subject to random disturbances where one employs measuring instruments to determine the state of the system; however, the measurements are often corrupted by noise due to the electrical and mechanical components of the measuring device. Finding the state of the system is called estimation and it should be stressed that estimation is essential for monitoring the system through its evolution. There is a vast literature on the subject and some recent books are: Gagic,[2] Cassidis and Junkms,[3] Fucks,[4] and Yedalli.[5] Some earlier engineering references are Aoki[6] and Jazwinski.[7] Beginning in the 1950s and continuing through the 1970s, the study of dynamic systems was gaining in popularity in the statistical literature, and one of the more influential papers was by Harrison and Stevens.[8]

One will find that Bayesian ideas have been important in the development of dynamic linear systems focusing on estimation, identification, and control. Indeed, the Kalman[9] filter is based on Bayesian inference and, as a result, many of the methodologies have a Bayesian foundation. The Kalman filter had a profound effect in electrical and mechanical engineering and will be explained in depth.

According to Petris, Petrone, and Campagnoli,[10, pp. 1-2] dynamic linear models were developed in the early sixties to control and monitor dynamic systems but early statistical contributions such as Thiele[11] go back to the nineteenth century. The main drive for the development of dynamic linear models is the advent of space travel with the NASA missions of Mercury and Apollo. This development expanded to many fields from biology to zoology.

Chapter 10 of Cowperwait and Metcalfe[12] gives a brief introduction to the dynamic linear models and will show the reader how to use R to generate data from such models and a maximum likelihood approach to estimation.

Another book from a Bayesian approach is that of West and Harrison[13] which generalizes the dynamic model to more complex observation and system equations. For example, the observation equation can easily include seasonal effects over time.

In what is to follow, the Bayesian posterior analysis will be developed for estimation, filtering, smoothing, and prediction, followed by many examples explained with the aid of R and WinBUGS.

8.2 Discrete Time Linear Dynamic Systems

Consider the model for the observation as

$$Y(t) = F(t)\theta(t) + U(t) \tag{8.1}$$

where

$Y(t)$ is an $n{\times}1$ observation vector, $F(t)$ an $n{\times}p$ nonrandom matrix, $\theta(t)$ a $p{\times}1$ unknown parameter vector, and $U(t)$ $n{\times}1$ observation error vector.

This model appears as a sequence of the usual type linear models, where $F(t)$ is the design matrix. The vector $\theta(t)$ is called the state of the system at time t and evolves according to the system equation

$$\theta(t) = G\theta(t-1) + V(t) \tag{8.2}$$

where

G is a $p \times p$ nonrandom transition matrix and $\theta(0)$ the initial state of the system. Also, the $p \times 1$ vector $V(t)$ is called the system error vector.

Some measuring device records the value $Y(t)$, which indirectly measures the value $\theta(t)$, and according to (8.1) is related in a linear way to $\theta(t)$ at time t. Since the signal $Y(t)$ is corrupted by noise, the additive error term $U(t)$ is added to the observation equation (8.1). Note that the system equation for the states of the system is represented by a first-order autoregressive process, but that higher order processes could have defined the system equation (8.2).

Based on the dynamic model, one is primarily interested in estimation, filtering smoothing, and prediction.

Each of these activities will be described in detail.

8.3 Estimation of the States

Estimation is estimating the states $\theta(t)$ of the system as observations $Y(1), Y(2), \ldots$ become available and each observation adds one parameter to be estimated. Suppose there are k observations $Y(1), Y(2), \ldots, Y(k)$, then there are k states $\theta(1), \theta(2), \ldots, \theta(k)$ to be estimated.

8.4 Filtering

Filtering is the estimation of the current state $\theta(k)$ which is executed with the Kalman filter. The Kalman filter is a recursive algorithm, thus, suppose $F(t)$ and G are known and that the initial state $\theta(0)$ has known mean $m(0)$ and known precision $p(0)$, and in addition suppose that $U(t)$ is normal with mean vector zero and precision matrix $p_u(t)$, and that $V(t)$ is distributed normal with mean vector zero and $p_v(t)$ as the precision matrix. Now assume that $\theta(0), U(1), U(2), \ldots, V(1), V(2), \ldots$ are independent, then given $Y(1), Y(2), \ldots, Y(k)$, the posterior distribution of $\theta(k)$ is normal with mean

$$m(k) = p^{-1}(k)\left\{F'(k)p_u(k)Y(k) + p_v(k)G[G'p_v(k)G + p(k-1)]^{-1}p(k-1)m(k-1)\right\}$$

$$(8.3)$$

and precision matrix

$$p(k) = F^{-1}(k)p_u(k)F(k) + [Gp_v^{-1}(k-1)G' + p_v^{-1}(k)]^{-1}. \tag{8.4}$$

Thus, the current state is recursively estimated by $m(k)$ and $p(k)$ expressed as a function of the current observation $Y(k)$, the current precision matrices $p_u(k)$ and $p_v(k)$, and the parameters $m(k-1)$ and $p(k-1)$ of the posterior distribution of the previous state $\theta(k-1)$. Since the posterior distribution is normal, the mean matrix and precision matrix are sufficient in the unique identification of the posterior distribution.

8.5 Smoothing

The problem of smoothing is: given the previous observations $Y(1), Y(2), \ldots, Y(k)$, how should the previous states $\theta(0), \theta(1), \ldots, \theta(k)$ be estimated? For the Bayesian, one would estimate them based on the posterior distribution of $\theta^{k-1} = [\theta(0), \theta(1), \ldots, \theta(k-1)]'$ which is derived from the posterior distribution of θ^k, which is the conditional distribution of $\theta^{(k)} = [\theta(0), \theta(1), \ldots, \theta(k)]'$, given $Y^k = [Y(1), Y(2), \ldots, Y(k)]'$, which implies all the previous are available to estimate the previous states. First, consider the posterior distribution of $\theta^{(k)}$ and the joint distribution of Y^k and $\theta^{(k)}$ which is given by

$$f(Y^k, \theta^{(k)}) = f_1[Y^k|\theta^{(k)}]f_2(\theta^{(k)}) \tag{8.5}$$

where $Y^k \in R^{nk}, \theta^{(k)} \in R^{p(k+1)}$ and

$$f_1[Y(k)|\theta^{(k)}] \propto \exp \sum_{t=1}^{t=k} [Y(t) - F(t)\theta(t)]'p_u(t)[Y(t) - F(t)\theta(t)] \tag{8.6}$$

and

$$f_2(\theta^{(k)}) = f_3(\theta^{(k)})f_4(\theta^{(k)}), \tag{8.7}$$

Also,

$$f_4(\theta^{(k)}) \propto \exp -(1/2) \sum_{t=1}^{t=k} [\theta(t) - G\theta(t-1)]'p_v(t)[\theta(t) - G\theta(t-1] \tag{8.8}$$

and

$$f_4(\theta^{(k)}) \propto \exp -(1/2)[\theta(0) - m(0)]'[\theta(0) - m(0)]. \tag{8.9}$$

Combining the exponents of f_1, f_2, the resulting exponent is quadratic in the $k+1$ states $\theta^{(k)}$ which implies the joint density (8.5) is a multivariate normal with density represented by

$$f(\theta^{(k)}, Y^k) \propto \exp -(1/2)[\theta^{(k)} - A^{-1}B]'A[\theta^{(k)} - A^{-1}B] \tag{8.10}$$

where A is of order $p(k+1)$ partitioned as

$$A = \begin{pmatrix} A_{11}, A_{12}, A_{13} \\ A_{21}, A_{22}, A_{23} \\ A_{31}, A_{32}, A_{33} \end{pmatrix}, \tag{8.11}$$

with $A_{13} = A_{31}$ $p \times p$ with all elements zero. In addition, it can be derived that A_{22} is quasi tridiagonal of order $(k-1)p$ and the i-th diagonal matrix is $F'(i)p_u(i)F(i) + p_v(i) + G'p_v(i+1)G, i = 1, 2, \ldots, k-1$. The super- and sub-diagonal matrices are $-G'p_u(i), i = 1, 2, \ldots, k-1$, while A_{23} is $(k-1)p$ by p, where

$$A_{23} = \begin{pmatrix} 0(pxp) \\ 0(pxp) \\ . \\ . \\ . \\ 0(pxp) \\ -Gp_v(k) \end{pmatrix} \tag{8.12}$$

and

$$B = \begin{pmatrix} p(0)m(0) \\ F_1'p_u(1)Y(1) \\ F_2'p_u(2)Y(2) \\ . \\ . \\ . \\ F_k'(k)p_u(k)Y(k) \end{pmatrix} \tag{8.13}$$

is $(k+1)p \times 1$. The scalar C is

$$C = \sum_{t=1}^{t=k} Y'(t)p_u(t)Y(t) + m'(0)p(0)m(0), \tag{8.14}$$

Therefore, the posterior distribution of $\theta^{(k)}$ is normal with mean vector $\mu = A^{-1}B$ and precision matrix $T = A$, which is denoted by

$$\theta^{(k)} \sim N_{k+1}(\mu, T). \tag{8.15}$$

How does one estimate the previous states? Note that the previous states are given by the $pk \times 1$ vector

$$\phi = [I, 0]\theta^{(k)} \tag{8.16}$$

where I is the identity matrix of order pk and 0 is the $pk \times p$ matrix of zeroes. The question is what is the posterior distribution of ϕ? It is obvious that the posterior distribution of ϕ is a multivariate normal with mean vector

$$E(\phi|data) = [I, 0]\mu \tag{8.17}$$

where

$$\mu = \begin{pmatrix} \mu_1 \\ \mu_2 \\ \cdot \\ \cdot \\ \mu_n \end{pmatrix} \tag{8.18}$$

and precision matrix

$$P(\phi|data) = \left\{ [I,0] T^{-1} [I,0]' \right\}^{-1} \tag{8.19}$$

where μ and T are given above.

Smoothing is a way to improve one's estimate of the previous states beyond that given by the Kalman filter.

8.6 Prediction

Suppose on the basis of k available observations $Y(1), Y(2), \ldots, Y(k)$, one wants to predict the future state $\theta(k+1)$, before observing $Y(k+1)$, then, the appropriate distribution with which to predict is the prior (prior to observing $Y(k+1)$) distribution of $\theta(k+1)$. The prior distribution of $\theta(k+1)$ is induced by the posterior distribution of $\theta(k)$, where

$$\theta(k+1) = G\theta(k) + V(k+1) \tag{8.20}$$

Remember the posterior distribution of $\theta^{(k)}$ is given by (8.15), with mean vector $m(k)$ and precision matrix $p(k)$ given by (8.3) and (8.4). Therefore, the prior distribution of $\theta(k+1)$ is easily found from (8.20). For example, the mean of this prior distribution is

$$m^*(k+1) = p^{*-1}(k+1)\left\{ p_v(k+1)G[G'p_v(k+1)G + p(k)]^{-1}p(k)m(k) \right\} \tag{8.21}$$

and corresponding precision matrix

$$p^*(k+1) = [Gp^{-1}(k)G' + p_v^{-1}(k+1)]^{-1}. \tag{8.22}$$

It should be emphasized $m^*(k+1)$ is a function of the first k observations because $m(k)$ is a function of those same observations. By repeated application of (8.21) and (8.22), one may predict $\theta(k+2)$ and $\theta(k+3)$, and so on, since the prior distribution of $\theta(k+2)$ is induced by that of $\theta(k+1)$ via the system equation

$$\theta(k+2) = G\theta(k+1) + V(k+2). \tag{8.23}$$

Thus, it follows that $\theta(k+2)$ is normal with mean vector

$$m^*(k+2) = p^{*-1}(k+2)\{p_v(k+2)G[G'p_v(k+2)G+ \\ p^*(k+1)]^{-1}p^*(k+1)m^*(k+1)\}. \tag{8.24}$$

and precision matrix

$$p^*(k+2) = [Gp^{*-1}(k+1)G' + p_v^{-1}(k+2)]^{-1}. \tag{8.25}$$

Prediction for dynamic linear models means something different than forecasting future observations, but instead refers to forecasting the future states of the system. As will be seen in the next section, the predictive distribution will play a crucial role for the control problem.

8.7 The Control Problem

The following discussion of the control problem is found in Broemeling,[1, pp. 244–254] but more recent information can be found in Petris, Petrone, and Campagnoli.[10, pp. 77–80] Consider an amended version of the dynamic linear model given by

$$Y(t) = F(t)\theta(t) + U(t) \tag{8.26}$$

$$\theta(t) = G\theta(t-1) + H(t)X(t-1) + V(t) \tag{8.27}$$

where the term $H(t)X(t-1)$ has been added to the system equation, where $H(t)$ is a known $p \times q$ matrix and
$X(t-1)$ is a $q \times 1$ vector of control variables.
The control problem is: How does one choose $X(0), X(1), \ldots$ in such a way that the corresponding states $\theta(1), \theta(2), \ldots$ are close to preassigned target values $T(1), T(2), \ldots$? Initially, one has yet to observe $Y(1)$ and $X(0)$ must be chosen in order that the state $\theta(1)$ be 'close' to $T(1)$, then having observed $Y(1)$ but not yet $Y(2)$, $X(1)$ is chosen so that $\theta(2)$ is close to $T(2)$, and so on. Choosing values for the control variables $X(0), X(1), \ldots$ is called a strategy, and one's strategy depends on how close is defined, that is the closeness of the states $\theta(i)$ to the corresponding targets $T(i), i = 1, 2, 3, \ldots$.

The strategy presented here is based on the predictive distribution, and is one of many strategies that could be used. See Aoki[6] and Maybeck[14] for description of other control strategies. The strategy chosen here is to choose control variables $X(0)$ and $X(1)$ that at the first stage

$$T(1) = E_1 E[\theta(1)|Y(1), X(0)] \tag{8.28}$$

and at the second
$$T(2) = E_2E[\theta(2)|Y(1), Y(2), X(0), X(1)], \tag{8.29}$$

and in general

$$T(i) = E_iE[\theta(i)|Y(1), Y(2), \ldots, Y(i-1)X(0), X(1), \ldots, X(i-1)], i = 1, 2, \ldots. \tag{8.30}$$

where E_i denotes expectation with respect to the conditional predictive distribution of $\theta(i)$ given $Y(j), j = 1, 2, \ldots, i-1$ and $X(i)$ is chosen in such a way that $T(i)$ is close to the posterior mean of $\theta(i)$ in the sense of (8.30). Thus, $X(0)$ is chosen so that the average value of the posterior mean of $\theta(1)$ is the target value $T(1)$ and the average is taken with respect to the predictive distribution of $Y(1)$, which has yet to be observed. At the second stage, $X(1)$ is chosen before $Y(2)$ is observed but after $Y(1)$ is observed, in such a way the target value $T(2)$ is the average value of the posterior mean of $\theta(2)$. Note that since $Y(1)$ has been observed and $Y(2)$ has yet to be observed, E_2 of (8.29) is with respect to the conditional predictive distribution of $Y(2)$ given the observed value of $Y(1)$, etc.

Let us return to the initial stage, $X(0)$ is chosen; thus, one must be able to find the posterior mean of $\theta(1)$, followed by the conditional predictive distribution of $Y(1)$. To do this, one must know the posterior distributions of $\theta(1)$ and that of the predictive distribution. Recall that the following distributions are assumed. Recall from the definition (8.1) and (8.2) of the dynamic linear model that $\theta(0) \sim N[m(0), p^{-1}(0)]$, where $P(0)$ is the precision matrix, $U(t) \sim N[0, p_u^{-1}(t)], V(t) \sim N[0, p_v^{-1}(t)]$, and $\theta(0), U(1), U(2), \ldots.V(1), V(2), \ldots..$ are independent. Under these assumptions, how should $X(0)$ be chosen?

First, it can be shown that the posterior mean of $\theta(1)$ is

$$E[\theta(1)|Y(1), X(0)] = p^{-1}(1)\{Gp^{-1}(0)G' + p_v^{-1}(1)]^{-1}H(1)X(0)+$$
$$F'(1)p_u(1)Y(1) + p_v(1)G[p(0) + G'p_v(1)G]^{-1}\}, \tag{8.31}$$

where the precision matrix

$$p(1) = F'(1)p_u(1)F(1) + [G'p^{-1}(0)G + p_v^{-1}(1)]^{-1}. \tag{8.32}$$

In addition, it can be shown that the predictive distribution of $Y(1)$ is normal with mean

$$E[Y(1)] = Q^{-1}(1)p_u(1)F'(1)p^{-1}(1)\{[Gp^{-1}(0)G' + p_v^{-1}(1)]^{-1}H(1)X(0)+$$
$$p_v(1)G[p(0) + G'p_v(1)G]^{-1}p(0)m(0)\}. \tag{8.33}$$

and precision matrix

$$Q(1) = p_u(1) - p_u(1)F(1)p^{-1}(1)F'(1)p_u(1). \tag{8.34}$$

In order to determine the value of the control, (8.31) and (8.33) are expressed as linear function of $X(0)$ and $Y(1)$.
It can be shown that

$$E[\theta(1)|X(0), Y(1)] = AX(0) + BY(1) + C, \tag{8.35}$$

where

$$A = p^{-1}(1)[Gp^{-1}(0)G' + p_v^{-1}(1)]^{-1}H(1),$$
$$B = p^{-1}(1)F'(1)p_u(1), \tag{8.36}$$

and

$$C = p^{-1}(1)p_v(1)G[p(0) + G'p_v(1)G]^{-1}p(0)m(0). \tag{8.37}$$

Also,

$$E[Y(1)|X(0), Y(1)] = DX(0) + E, \tag{8.38}$$

where

$$D = Q^{-1}(1)p_u(1)F'(1)p^{-1}(1)[Gp^{-1}(0)G' + p_v^{-1}(1)]^{-1}H(1), \tag{8.39}$$

and

$$E = Q^{-1}(1)p_u(1)F'(1)p^{-1}(1)p_v(1)G[p(0) + G'p_v(1)G]^{-1}p(0)m(0). \tag{8.40}$$

Thus, let

$$T(1) = (A + BD)X(0) + BE + C$$
and set the value of the control variable at \tag{8.41}

$$X(\tilde{0}) = \{(A + BD)'(A + BD)]^{-1}\}(A + BD)'[T(1) - BE - C], \tag{8.42}$$

which works for the first stage, but what about the general solution for the control problem?
In the general case, one needs the conditional predictive distribution of $Y(i)$ given $Y(j), j = 1, 2, \ldots.i = 1$ and the posterior mean of $\theta(i)$, then solve Equation (8.30) for $X(i-1)$.
The control strategy is summarized as follows:
Consider the linear dynamic system (8.1) and (8.20), where
$\theta(0) \sim N[m(0), p^{-1}(0)], U(t) \sim N[0, p_u^{-1}(t)], V(t) \sim N[0, p_v^{-1}(t)], t = 1, 2, \ldots$
and that
$G, F(t), H(t), p_u(t), p_v(t), m(0)$, and $p(0)$ are known. Also, assume that $\theta(0); U(1), U(2), \ldots, V(1), V(2), \ldots$

are stochastically independent, then at the i-th stage of control, the value of $X(i-1)$, say $X(i-1)$

$$T(i) = E_i E[\theta(i)|Y(1), Y(2), \ldots, Y(i-1)X(\widetilde{0}), X(\widetilde{1}), \ldots, X(i-1)] \qquad (8.43)$$

Therefore, the general solution is

$$X(\widetilde{i}-1) = \{[A(i) + B(i)D(i)]'[A(i) + B(i)D(i)]\}^{-1}, \qquad (8.44)$$

where

$$A(i) = p^{-1}(i)[Gp^{-1}(i-1)G' + p_v^{-1}(i)]^{-1}H(i), \qquad (8.45)$$

$$\begin{aligned} B(i) &= p^{-1}(i)F'(i)F(i)p_u(i), \\ C(i) &= p^{-1}(i)p_v(i)G[p(i-1) + G'p_v(i)G]^{-1}p(i-1)m(i-1), \end{aligned} \qquad (8.46)$$

$$\begin{aligned} p(i) &= F'(i)p_u(i)F(i) + [Gp^{-1}(i-1)G + p_v^{-1}(i)]^{-1}, \\ D(i) &= Q^{-1}(i)p_u(i)F(i)p^{-1}(i)[Gp^{-1}(i-1)G' + p_v^{-1}(i)]^{-1}H(i), \end{aligned} \qquad (8.47)$$

$$\begin{aligned} E(i) &= Q^{-1}(i)p_u(i)F(i)p^{-1}(i)p_v(i)G[p(i-1) + G'p_v(i)G]^{-1}p(i-1)m(i-1), \\ Q(i) &= p_u(i) - p_u(i)F(i)p^{-1}(i)F'(i)p_u(i), \end{aligned}$$

$$\qquad (8.48)$$

and

$$\begin{aligned} m(i) = {}&p^{-1}(i)\{[Gp^{-1}(i-1)G' + p_v^{-1}(i)]^{-1}H(i)X(i-1) + F'(i)p_u(i)Y(i) + \\ &p_v(i)G[p(i-1) + G'p_v(i)G]^{-1}p(i-1)m(i-1). \end{aligned}$$

Thus, one can conclude that $\theta(i) \sim N[m(i), p^{-1}(i)], i = 1, 2, \ldots$, where $m(i)$ is the mean vector and $p(i)$ the precision matrix.

What is the predictive distribution of $Y(i)$ given $Y(j), j = 1, 2, \ldots, i-1$? It can be shown that it has a normal distribution with mean

$$E[Y(i)|Y(j) = y(j), j = 1, 2, \ldots, i-1] = Q(i)R(i), \qquad (8.49)$$

where $Q(i)$ is given by (8.48) and

$$R(i) = p_u(i)F'(i)p^{-1}(i)\{[Gp^{-1}(i-1)G' + p_v^{-1}(i)]^{-1}H(i)X(i-1) + R_1(i), \qquad (8.50)$$

and

$$R_1(i) = p_v(i)G[p(i-1) + G'p_v(i)G]^{-1}p(i-1)m(i-1). \qquad (8.51)$$

See Broemeling[1, pp. 244–251] for additional details on the explanation for the control strategy of linear dynamic systems.

Since $m(i)$ is a linear function of $Y(i)$ and $X(i-1)$ and because the mean (8.49) is a linear function of $x(i-1)$, the values $X(i)$ that control the system are easily computed via (8.44). The foregoing section shows how one can compute the control values $X(i), i = 0, 1, \ldots$ in order to control the system (states of the system) at target values $T(i), i = 1, 2, \ldots$ in the sense that the posterior mean of $\theta(i), i = 1, 2, \ldots$ is the desired target value.

8.7.1 Example: The Kalman Filter

The preceding equations introduce to the reader the Bayesian analysis of linear dynamic models, including that for estimation, filtering, smoothing, prediction, and control strategies. These methods will be illustrated with the dynamic model

$$Y(t) = 3\theta(t) + U(t), t = 1, 2, \ldots 50$$
$$\theta(t) = 0.5\theta(t-1) + V(t) \tag{8.52}$$

where $U(t) \sim N(0, .2), V(t) \sim N(0, 1), \theta(0) \sim N(2, .7)$, and $\theta(0); U(1), U(2), \ldots, U(50); V(1), V(2), \ldots, V(50)$ are jointly independent. In terms of the general model, $F(t) = 3$, $G = 5$, $P_u(t) = 5, P_v(t) = 1, m(0) = .2$, and $p(0) = 1/.7$.

Fifty observations are generated from the model (8.52) and the results are reported in Table 8.1 which contains six columns: (1) the U column for the observations errors, (2) the system equation errors, (3) the states of the system beginning with the initial value in the first row, the Kalman filter value, and (5) the Y column for the observations, and (6) the precision of the posterior distribution of the current state of the system. Note this is a Bayesian posterior analysis initiated with the prior distribution $\theta(0) \sim N(0, .7)$, and the table values were generated with R.

Since the design matrices and the two precision matrices are constants, the precision of the estimates reach the level 18.987 just after three iterations and remains at the level for the remainder of the 40 observations. It is also seen the Kalman filter estimate imitates the behavior of the states with regard to the sign of the estimate and actual state value.

With R, 100 observations and states are generated according to the model (8.52)

The R statements are given by

RC 8.1
```
> y<-x<-u<-v<-rnorm(100)
> for ( t in 2:100) y[t]<-3*x[t]+u[t]
> for ( t in 2:100) x[t]<-0.5*x[t-1]+v[t]
```

TABLE 8.1

The Dynamic Linear Model (8.52)

Row	U	V	θ	m	Y	P
1	.3826	.1421	2.9944	2.000	0	.7000
2	−1.0744	1.2243	2.7215	.2744	7.09	18.7368
3	.1626	−.7446	.6141	.6927	2.005	18.9868
4	−.6353	.7390	1.0451	.8089	2.503	18.9870
5	.8767	.3159	.8390	1.0934	3.394	18.897
6	−.6793	−.256	.1635	−.0312	−.189	18.987
7	.8093	−1.092	−1.0103	−.7029	−2.222	18.987
8	−1.0961	−.1825	−.6876	−1.016	−3.159	18.987
9	1.485	1.697	1.2637	1.6408	5.276	18.987
10	1.0534	−.4624	.1694	.5362	1.562	18.987
11	−1.228	1.992	2.0774	1.5952	5.004	18.987
12	−2.605	−1.517	−.4788	−1.235	−4.042	18.987
13	−.148	−3.3883	−3.5977	−3.489	−10.492	18.987
14	−2.063	.2301	−1.568	−2.23	−6.77	18.987
15	−.8232	2.132	1.348	.960	3.221	18.987
16	1.376	.3588	1.032	1.439	4.475	18.987
17	−.3125	.0436	.560	.469	1.368	18.987
18	−2.075	−1.5534	−1.2734	−1.85	−5.89	18.987
19	1.494	−1.047	−1.68	−1.172	−3.558	18.989
20	1.834	.0716	−.7704	−.1811	−.477	18.987
21	−.95	.438	.052	−.254	−.792	18.987
22	1.902	−.188	−.161	.4414	1.418	18.987
23	−1.87	.078	−.002	−.584	−1.887	18.987
24	−1.761	−.754	−.755	−1.288	−4.028	18.987
25	−1.17	−.754	−.755	−1.288	−4.028	18.987
26	−.56	.951	1.487	1.249	3.09	18.987

(Continued)

TABLE 8.1 (Cont.)

Row	U	V	θ	m	Y	P
27	2.671	1.133	1.877	2.656	8.303	18.987
28	−2.87	.506	1.445	.5319	1.465	18.987
29	−1.52	1.705	2.427	1.832	5.755	18.987
30	1.319	.489	1.703	2.079	6.42	18.987
31	−.942	−.256	.595	.321	.845	18.987
32	−.609	.478	.776	.551	1.72	18.987
33	.2991	.827	1.216	1.261	3.947	18.987
34	−.059	.051	.659	.639	1.919	18.987
35	.891	.534	.864	1.118	3.485	18.987
36	1.663	−1.213	−.78	−.185	−.67	18.987
37	.945	−1.556	−1.947	−1.552	−4.89	18.987
38	−.984	1.262	.289	−.077	−.117	18.987
39	.571	.130	.275	.439	1.397	18.987
40	2.004	.057	.194	.829	2.58	18.987
41	1.29	.176	.273	.689	2.114	18.987
42	−.572	−1.25	−1.115	−1.22	−3.91	18.987
43	.395	.156	−.4008	−.286	−.807	18.987
44	2.709	−.086	−.287	.576	1.848	18.987
45	−2.013	−.348	−.491	−1.087	−3.489	18.987
46	.647	.131	−.114	.067	.304	18.987
47	2.662	−2.76	−2.818	−1.83	−5.79	18.987
48	−.992	−1.018	−2.428	−2.663	−8.278	18.987
49	.391	1.494	.2808	.3206	1.234	18.987
50	2.143	1.039	1.180	1.8043	5.683	18.987
avg	−.2302	.9998	1.589	1.481	4.539	18.987

and the 100 responses are:
Y =(0.14959198, −5.37012593, 2.21321230, 6 .365985137, −2.34751838,
−7.32950922, 3.55255773, 6.37395389, 2.06741868, −5.18268672, 0.21846230,
−3.13859749, −4.19741128, 9.32204785, 5.61082153, 3.77040340, 3.30503315,
−3.24616196, 1.90499312, 4.08503363, 2.58153230, 4.17257421, −1.21747645,
9.90844367, 3.88488269, 7.46839674, 2.68816988, −1.23181352, 2.14609487,
3.29948026, −3.85560592, −3.42033004, 7.54778777, −1.56727749,
−3.92253180, 2.74932840, −2.02017407, 8.63087927, −2.39919026, −2.77818677,
0.89570163, −4.62489332, 1.68967411, −5.29902102, 0.56433725,
−2.14419199, −1.24642434, 6.22443857, −1.79213317, 1.28449415,
−4.92068899, −5.29623477, 5.04496910, 5.27692688, −0.32301505,
−2.02035923, −0.20861437, 2.51544250, 8.72000959, −0.27606923, 6.17945441,
5.28580807, 1.28860630, 6.12382048, −1.68495877, −4.63528409,
−7.38147316, 4.62930115, −8.49419952, −4.78412607, 6.56876796, 3.53461933,
2.09950356, −4.73863626, 10.62315309, −4.19165484, −4.04449009, 2.67568661,
0.51670917, −1.69030747, −4.56105658, −5.17486116, −2.37879508,
−6.00325632, 0.06342277, 2.16067827, −6.18916787, 3.39861172, 3.58405274,
0.55476399, −6.47731329, 2.19359169, 0.78112862, −3.22599198,
−0.43449694, −1.00378650, 6.79738667, −1.37719519, 0.27108825, −2.60227891)

The plot of the responses versus time is portrayed in Figure 8.1.
Below is the WinBugs code for the posterior analysis of the dynamic
linear model (8.52). The data used for the analysis was generated by the
R commands of RC 8.1 using the following values for the parameters:
$\beta = 3$, $\delta = .5$, and $\tau_1 = 1 = \tau_2$. The simulation is based on 45,000 iterations
with 5,000 initial iterations.

BC 8.1
model

```
{
#dlm model

# The following is the code for the system equation. See (8.52)
mu[1] <- delta*y0
theta[1] ~ dnorm(mu[1],tau2)
for(t in 2:100) {
mu[t] <- delta*theta[t-1]
theta[t] ~ dnorm(mu[t],tau2)}
tau2~dgamma(.1,.1)
y0<-0
delta~dnorm(.5,1)

# The following is the code for the observation equation. See (8.52)
```

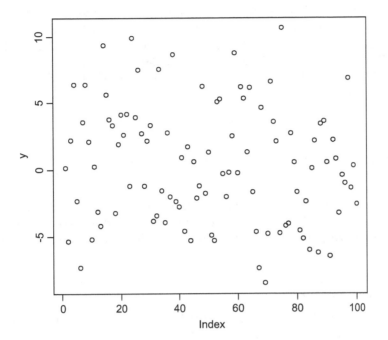

FIGURE 8.1
Responses for Dynamic Linear Model (8.52)

```
for(t in 1:100){v[t]<-beta*theta[t]
y[t]~dnorm(v[t],tau1)}
tau1~dgamma(.1,.1)
beta~dnorm(3,1)
}
```

list(y=c(0.14959198, −5.37012593, 2.21321230, 6.365985137, −2.34751838, −7.32950922, 3.55255773, 6.37395389, 2.06741868, −5.18268672, 0.21846230, −3.13859749, −4.19741128, 9.32204785, 5.61082153, 3.77040340, 3.30503315, −3.24616196, 1.90499312, 4.08503363, 2.58153230, 4.17257421, −1.21747645, 9.90844367, 3.88488269, 7.46839674, 2.68816988, −1.23181352, 2.14609487, 3.29948026, −3.85560592, −3.42033004, 7.54778777, −1.56727749, −3.92253180, 2.74932840, −2.02017407, 8.63087927, −2.39919026, −2.77818677, 0.89570163, −4.62489332, 1.68967411, −5.29902102, 0.56433725, −2.14419199, −1.24642434, 6.22443857, −1.79213317, 1.28449415, −4.92068899, −5.29623477, 5.04496910, 5.27692688, −0.32301505, −2.02035923, −0.20861437, 2.51544250, 8.72000959, −0.27606923, 6.17945441, 5.28580807, 1.28860630, 6.12382048, −1.68495877, −4.63528409, −7.38147316, 4.62930115, −8.49419952, −4.78412607, 6.56876796, 3.53461933, 2.09950356,

–4.73863626,   10.62315309,   –4.19165484,   –4.04449009,   2.67568661,
0.51670917, –1.69030747, –4.56105658, –5.17486116, –2.37879508,
–6.00325632,   0.06342277, 2.16067827, –6.18916787, 3.39861172, 3.58405274,
0.55476399, –6.47731329, 2.19359169, 0.78112862, –3.22599198,
–0.43449694, –1.00378650, 6.79738667, –1.37719519, 0.27108825,
–2.60227891))

list(beta=3,delta=.5,tau1=1,tau2=1)

The posterior analysis is reported in Table 8.2.

The second column is the value of the parameter used to generate the 100 data values via RC 8.1.

It is seen that the posterior mean and median of β is quite close to the 'true' value of 3. The same is not true for the other posterior distributions; however, the 95% credible intervals do indeed contain those true values.

8.7.2 The Control Problem

The control problem is now considered via the following equations for the observation and state of the system, namely

$$
\begin{aligned}
Y(t) &= 3\theta(t) + U(t) \\
\theta(t) &= 0.5\theta(t-1) + 2X(t-1) + V(t)
\end{aligned}
\tag{8.53}
$$

where

$Y(t), \theta(t), U(t), V(t)$ are scalar, and $\theta(0) \sim N(.2, 1.428)$. In addition, $U(t) \sim N(0, p_u^{-1}), V(t) \sim N(0, p_v^{-1})$. Suppose the states of the system are to be controlled at the target value $T(t) = -.7, t = 1, 2, \ldots, 39$ in such a way the future mean of each state equals the target value. Using formula (8.44), forty observations are generated where $p_U = p_V = 1$. The observations are generated as follows:

TABLE 8.2

Posterior Analysis for Dynamic Linear Model

Parameter	Value	Mean	SD	Error	2½	Median	97½
β	3	2.768	1.08	.037	.5743	2.772	4.822
δ	.5	.0903	.3195	.0082	-.5434	.0522	.7943
τ_1	1	.8535	2.844	.1111	.0425	.0699	7.414
τ_2	1	3.855	5.703	.2147	.0896	1.623	20.55

(a) The initial state is selected so that $\theta(0) \sim N(.2, 1.482)$.

(b) $X(0)$ is computed with (8.44).

(c) $V(1)$ is selected according to $V(1) \sim N(0, p_V^{-1})$

(d) $\theta(1)$ is computed via (8.53).

(e) $U(1)$ is selected according to $U(1) \sim N(0, p_U^{-1})$

(f) $Y(1)$ is selected via (8.53)

(g) The first six steps are repeated 39 times generating the observations and values for the control variable x.

Recall that the control strategy is to choose the control variable in such a way that the mean of the posterior distribution of the state is set equal to the target value.

See Table 8.3 for the results of this experiment.

How well does this strategy work? Is the average of the posterior distribution of the state come close to the target of $T=.7$? The student will be asked these questions as exercises at the end of the chapter. The value of the Kalman filter is graphed below as a function of time varying over the times 1,2, ...,40. See Figure 8.2.

8.8 Adaptive Estimation

The Kalman filter was designed for linear dynamic systems where all the parameters are known, except the state parameters. The parameters of a linear dynamic models are $m(0), p(0)$, $p_U(t), p_V(t)$, where $m(0)$ and $p(0)$ are the mean and precision of the initial state $\theta(t)$, $p_U(t)$ the precision of the observations equation, and $p_V(t)$ the system equation errors. Of course, in practice, one or more of the parameters are unknown and such situations are referred to as adaptive estimation. Adaptive estimation in the area of dynamic linear models was and is an active area of research. For example, see Broemeling,[1, pp. 274–304] Jazwinski,[7] and Maybeck[14] for some early contributions and Steigerwald[15] for a more recent paper. Adaptive filtering based on Bayesian methods is provided by Alspach[16] and Hawkes.[17] See Broemeling[1, pp. 305–306] for a more extensive list relevant to adaptive estimation.

The approach taken here is to first consider the special case of when the observation and system equations are univariate with unknown precisions for the observation and system equations. Thus, consider the dynamic system.

TABLE 8.3

The Control of a Dynamic System at $T=-.7$

Time	P	Q	A	B	C	D	E	Z	X	Y	m
1	.70000	.00000	.00000	.00000	.00000	.00000	.00000	-.400	.732500	.00000	.200000
2	.7368	.0576	.1513	.3081	.0075	6	.30000	-.313	-.0426	-.298	-.14480
3	.9749	.0977	.1954	.3007	-.0070	6	-.2172	-.226	-.2333	-1.416	-.4924
4	9.975	.0977	.1955	.3007	-.0240	6	-.7387	-.123	-.3283	-2.878	-.9068
5	9.975	.0977	.1955	.3007	-.0443	6	-1.360	-.315	-.3921	-.234	-.1389
6	9.975	.0977	.1955	.3007	-.0067	6	-.2084	-.235	-.2221	-1.291	-.4767
7	9.975	.0977	.1955	.3007	-.0223	6	-.685	-.014	-1.263	-4.229	-1.3404
8	9.975	.0977	.1955	.3007	-.065	6	-2.010	-.304	.0335	-.380	-.1828
9	9.975	.0977	.1955	.3007	-.0089	6	-.2743	.0521	-1.3693	-5.121	-1.6086
10	9.975	.0977	.1955	.3007	-.0786	6	-2.412	.2887	-2.4751	-8.268	-2.55
11	9.975	.0977	.1955	.3007	-.1249	6	-3.832	.1574	-2.653	-6.521	-2.029
12	9.975	.0977	.1955	.3007	-.0999	6	-3.044	.1507	-2.209	-6.433	-2.003
13	9.975	.0977	.1955	.3007	-.097	6	-3.004	.0371	-1.476	-4.922	-1.548
14	9.975	.0977	.1955	.3007	-.075	6	-2.332	-.355	-.535	.303	.0228
15	9.975	.0977	.1955	.3007	.001	6	.034	-.096	-.767	-3.143	-1.013
16	9.975	.0977	.1955	.3007	-.049	6	-1.520	-.154	-.778	-2.374	-.782

17	9.975	.0977	.1955	.3007	−.038	6	−1.173	−.471	.3691	1.843	.485
18	9.975	.0977	.1955	.3007	.023	6	.728	−.168	−.851	−2.185	−.725
19	9.975	.0977	.1955	.3007	−.035	6	−1.088	−.308	−.093	−.324	−.165
20	9.975	.0977	.1955	.3007	−.008	6	−.248	−.506	1.255	2.312	.626
21	9.975	.0977	.1955	.3007	.030	6	.9401	−.203	−.2098	−1.721	−.5859
22	9.975	.0977	.1955	.3007	−.028	6	−.878	.060	−1.835	−5.329	−1.643
23	9.975	.0977	.1955	.3007	−.080	6	−2.46	−.259	−.681	−.982	−.363
24	9.975	.0977	.1955	.3007	−.0177	6	−.545	−.355	−.462	.302	.022
25	9.975	.0977	.1955	.3007	.0011	6	.0337	−.302	−.553	−.401	−.188
26	9.975	.0977	.1955	.3007	−.009	6	−.283	.441	−3.211	−10.29	−3.164
27	9.975	.0977	.1955	.3007	−.154	6	−4.746	.014	−1.444	−4.616	−1.456
28	9.975	.0977	.1955	.3007	0.071	6	−2.185	−.441	.7725	1.443	.365
29	9.975	.0977	.1955	.3007	.017	6	.548	−.322	−.056	−.145	−.112
30	9.975	.0977	.1955	.3007	−.005	6	−.168	−.308	−.547	−.320	−.164
31	9.975	.0977	.1955	.3007	−.008	6	−.247	−.421	.0093	1.176	.285
32	9.975	.0977	.1955	.3007	.013	6	.427	−.194	−.475	−1.84	−.621
33	9.975	.0977	.1955	.3007	−.030	6	−.932	−.238	−.562	−1.25	−.445
34	9.975	.0977	.1955	.3007	−.021	6	−.668	.307	−2.486	−8.512	−2.628

(Continued)

TABLE 8.3 (Cont.)

Time	P	Q	A	B	C	D	E	Z	X	Y	m
35	9.975	.0977	.1955	.3007	-.128	6	-3.942	-.170	-.917	-2.158	-.717
35	9.975	.0977	.1955	.3007	-.035	6	-1.076	-.134	-.480	-2.633	-.860
37	9.975	.0977	.1955	.3007	-.042	6	-1.290	-.797	1.485	6.182	1.790
38	9.975	.0977	.1955	.3007	.087	6	2.685	-.064	-1.504	-3.566	-1.14
39	9.975	.0977	.1955	.3007	-.055	6	-1.711	.570	-3.435	-12.01	-3.682
40	9.975	.0977	.1955	.3007	-.180	6	-5.523	-.521	.9957	2.511	.6865

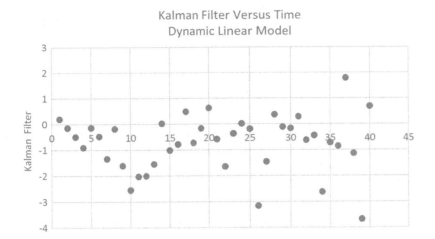

FIGURE 8.2
The Kalman Filter Values Dynamic Linear Model (8.53)

$$Y(t) = F(t)\theta(t) + U(t)$$
$$\theta(t) = G\theta(t-1) + V(t)$$

(8.54)

where $U(t) \sim N(0, \tau_U^{-1})$, $V(t) \sim N(0, \tau_V^{-1})$, and τ_U and τ_V are unknown positive precision parameters of the observation and system equations, respectively. Also, the observation error sequence $U(t)$, $t = 1, 2, \ldots$ is a sequence of independent random variables and independent of the sequence of systems errors $V(t)$, $t = 1, 2, \ldots$ The other parameters $m(0)$ and $p(0)$ are the known mean and precision of the initial state $\theta(0)$. As observations become available, the corresponding states are to be estimated along with the two precision parameters τ_U and τ_V. Recall with the Kalman filter, the states were estimated in a iterative fashion, where the prior of the current state induced a posterior distribution for the next future state. The precision parameters are assigned prior distributions as follows. Let the prior density of τ_U be gamma with parameters α_u and β_u, namely

$$g(\tau_U) \propto \tau_U^{\alpha_u - 1} e^{-\tau_U \beta_u}, \tau_U > 0$$

(8.55)

and in a similar manner, let the density of τ_V be

$$g(\tau_V) \propto \tau_V^{\alpha_v - 1} e^{-\tau_V \beta_v}, \tau_V > 0$$

(8.56)

that is gamma with parameters α_v and β_v.

First, consider estimating $\theta(1)$, then it can be shown that its posterior density is

$$g[\theta(1)|y(1)] \propto \{2\beta_u + F^2(1)[y(1)/F(1) - \theta(1)]^2\}^{-(2\alpha_u+1)/2}$$
$$X\{2\beta_v + [\theta(1) - G\theta(0)]^2\}^{-(2\alpha_v+1)/2}, \theta(1) \in R$$

which is a 2/0 poly-t density and is well known in econometrics. For example, see Dreze[18] for a review of the literature on this topic. In general, it can be shown that $\theta^{(k)} = [\theta(1), \theta(2), \ldots, \theta(k)]'$ has the posterior density

$$g(\theta^{(k)})|y(1), y(2), \ldots, y(k)] \propto \{2\beta_u + \sum_{t=1}^{t=k} F^2(t)[y(t)/F(t) - \theta(t)]^2\}^{(h+2\alpha_u)/2}$$

$$(8.57)$$

$$X\{2\beta_v + \sum_{t=1}^{t=k} [\theta(t) - G\theta(t-1)]^2\}^{(k+2\alpha_v)/2}$$

where $F(t)$ and G are given by (8.54), and $\theta^{(k)} \in R^k$. This density is that of a 2/0 k variable poly-t distribution and is very difficult to work with. For example, the normalizing constant is unknown, and thus the constant and the moments of the distribution must be determined numerically. Various approximations have been proposed for this distribution. For example, Box and Tiao[19] and Zellner[20] have proposed infinite series while Dreze[18] had developed numerical integrations to find the moments and other characteristics of the distribution.

Is it possible to find a recursive estimate like the Kalman filter to estimate the successive states $\theta(t)$? One could approximate each t density factors by a normal distribution with the same mean and variance, then the posterior density of $\theta^{(k)}$ is normal and it would be possible to find a Kalman-type filter expression for the posterior mean of $\theta(k)$. However, it should be noted that the normal approximation might be inadequate leading to erroneous estimates of the states; thus, this approach will not be taken up. Let us try another approach by considering the conditional posterior density of $\theta(k)$ given $\theta(t) = \tilde{\theta}(t), t = 1, 2, \ldots, k-1$, which is a 2/0 univariate poly-t with density

$$g[\theta(k)|\theta(t) = \tilde{\theta}(t), t = 1, 2, \ldots, k-1] = g_1[\theta(k)|\theta(t) = \tilde{\theta}(t), t = 1, 2, \ldots, k-1]$$
$$X g_2[\theta(k)|\theta(t) = \tilde{\theta}(t), t = 1, 2, \ldots, k-1]$$

where

$$g_1[\theta(k)|\theta(t) = \theta(\widetilde{t}), t = 1, 2, \ldots, k-1] \propto \{1 + N_1/N_2\}^{-(k+2\alpha_u)/2}, \qquad (8.58)$$

with

$$N_1 = F^2(k)[\theta(k) - y(k)/F(k)]^2, \qquad (8.59)$$

and

$$N_2 = 2\beta_u + \sum_{t=1}^{t=k-1} F_2(t)[(y(t)/F(t) - \theta(\widetilde{t})]^2. \qquad (8.60)$$

Also,

$$g_2[\theta(k)|\theta(t) = \theta(\widetilde{t}), t = 1, 2, \ldots, k-1] \propto \{1 + N_3/N_4\}^{-(k+2\alpha_v)/2}. \qquad (8.61)$$

where

$$N_3 = [\theta(k) - G\theta^\sim(k-1))]^2, \qquad (8.62)$$

and

$$N_4 = 2\beta_v + \sum_{t=1}^{t=k-1} [\theta^\sim(t) - G^\sim\theta(t-1)]^2. \qquad (8.63)$$

The first factor g_1 is a t density with $k \mid 2\alpha_u - 1$ degrees of freedom, location $y(k)/F(k)$, and precision

$$p_1(k) = (k + 2\alpha_u - 1)F^2(k)/\left\{2\beta_u + \sum_{t=1}^{t=k-1} F^2(t)[y(t)/F(t) - \theta(\widetilde{t})]^2\right\} \qquad (8.64)$$

while the degrees of freedom for the second t density is $k + 2\alpha_v - 1$, location $G\theta^\sim(k-1)$, and precision

$$p_2(k) = (k + 2\alpha_v - 1)/\{2\beta_v + \sum_{t=1}^{t=k-1} [\theta^\sim(t) - G\theta^\sim(t-1)]^2 \qquad (8.65)$$

Since these two densities are univariate, it is easy to numerically find the mean, mode, variance, and other conditional characteristics of the distribution. A recursive estimate of $\theta(k)$ is found as follows: for $k=1$, the conditional density of $\theta(1)$ given $\theta(0) = \theta(\widetilde{0})$ is a 2/0 poly-t given by (8.58) and (8.61) and the mode of this density is easily calculated as $\theta(\widetilde{1})$. At the second stage, the mode of the conditional density of $\theta(2)$ given $\theta(1) = \theta(\widetilde{1})$ is computed to be $\theta(\widetilde{2})$, where the first mode $\theta(\widetilde{1})$ depends on $y(1)$, but the second mode $\theta(\widetilde{2})$ and depends on $y(1)$ and $y(2)$, etc. The iteration is continued, and it is seen that the mode $\widetilde{\theta(k)}$ of conditional distribution of $\theta(k)$, given $\theta(t) = \theta^{\sim}(t)$, $t = 1, 2, \ldots, k-1$ depends on $y(1), y(2), \ldots, y(k)$. Thus, it is seen as a recursive estimate of $\theta(k)$ is easily given by the sequence of conditional 2/0 poly-t distributions. Instead of the sequence of conditional modes, one could compute a corresponding sequence of conditional means

$$\theta^*(t) = E[\theta(t)| \theta^*(1), \theta^*(2), \ldots, \theta^*(t-1)] \tag{8.66}$$

where $\theta^*(1)$ is the conditional mean of $\theta(1)$ given $\theta(0)$ and $y(1)$.

The general case is now considered, where the dynamic linear model is given by

$$\begin{aligned} Y(t) &= F(t)\theta(t) + u(t) \\ \theta(t) &= G\theta(t-1) + v(t) \end{aligned} \tag{8.67}$$

for $t=1,2,\ldots$, where the observation vector $Y(t)$ is n by 1, $\theta(t)$ the state at time t is p by 1, $F(t)$ is a known n by p design matrix, G is a $p \times p$ known system matrix, and the $n \times 1$ vector observation errors are

$$u(t) \sim nid(0, p_u^{-1}), \tag{8.68}$$

while the systems vector errors are

$$v(t) \sim nid(0, p_v^{-1}). \tag{8.69}$$

p_u is a $n \times n$ unknown precision matrix for the observation equation and p_v the p-th order unknown precision matrix for the system equation.

It is assumed that $\theta(0)$ is known and that the independent sequence of observations errors $u(t)$ is independent of the independent sequence of system errors $v(t)$, $t=1,2,\ldots$.

Suppose a priori that the precision matrices p_u and p_v have independent Wishart distributions, then it is possible to derive a recursive estimator of the states $\theta(t), t = 1, 2, \ldots$.

8.9 An Example of Adaptive Estimation

This example will depict adaptive filtering of the bivariate states of a linear dynamic model with fifty bivariate observations according to the model

$$Y(t) = F\theta(t) + U(t)$$
$$\theta(t) = G\theta(t-1) + V(t) \tag{8.70}$$

where $t=1,2,\ldots,50$, and

$$F = \begin{pmatrix} 3.5, 3 \\ 3, 3.0 \end{pmatrix}, \tag{8.71}$$

$$G = \begin{pmatrix} .5, .5 \\ .5, .5 \end{pmatrix}, \tag{8.72}$$

$U(t) \sim BVN(0, p_u^{-1})$, $V(t) \sim BVN(0, p_v^{-1})$, and $\theta(0)$ is known. Also, assume the $U(1), U(2), \ldots, U(50)$ are independent of the sequence $V(1), V(2), \ldots, V(50)$. Note the observation precision matrix is p_u and that for the system equation is p_v.

The fifty observations are generated according to (8.70): $Y(1)$ was generated by selecting $V(1)$ from a $BVN(0, p_v^{-1})$ population, and then $\theta(1)$ generated according to the system model $\theta(1) = G\theta(0) + V(1)$, which was followed by $U(1)$ being selected from the $BVN(0, p_u^{-1})$ population followed by $Y(1)$ generated according to the observation equation $Y(1) = F\theta(1) + U(1)$, etc. Table 8.4 consists of four columns, where the first two consist of the bivariate system vectors $\theta(1), \theta(2), \ldots, \theta(50)$ which are the fifty corresponding filter estimates according to (8.66).

The precision matrix for the observation equation is given by

$$p_u = \begin{pmatrix} .6, -.4 \\ -.4, .6 \end{pmatrix}, \tag{8.73}$$

and that for the system equation by

$$p_v = \begin{pmatrix} 555.55, -444.44 \\ -444.44, 555.55 \end{pmatrix}. \tag{8.74}$$

Finally, the initial state is given by the vector value

$$\theta(0) = \begin{pmatrix} 1 \\ 1 \end{pmatrix}.$$

With regard to prior information, the two precision matrices are given independent Wishart distributions, where p_u has 4 degrees of freedom and precision matrix T_u and p_v has 4 degrees of freedom and precision matrix T_v. Also, it should be noted that $T_u = 4p_u^{-1}$ and $T_v = 4p_v^{-1}$.

Note that the first two columns are the values of two states at time t, namely

TABLE 8.4

Bivariate Observations for Adaptive Estimation

1	.8767	.9436	.9921	.9921
2	1.0138	1.0086	.9956	.9949
3	1.0157	.9275	.9968	.9964
4	.8869	.8818	.9959	.9950
5	.9979	.9660	.9960	.9958
6	1.0177	.9694	.9971	.9968
7	.8915	.9432	.9970	.9969
8	.9890	1.0239	.9964	.9968
9	1.045	1.0531	.9965	.9969
10	.8683	.8866	.9977	.9972
11	1.0401	1.1292	.9984	.9980
12	.9417	.9990	.9983	.9984
13	1.0145	.9442	.9992	.9993
14	.9947	1.0194	.9993	.9996
15	1.04002	1.06398	.99972	.99996
16	1.09811	1.00447	.99955	.99921
17	1.11507	1.06270	.99892	.99972
18	.89765	.79794	.99937	.99956
19	.95830	.97013	.99926	.99943
20	.96068	.85402	.99924	.99913
21	1.06411	1.05442	.99890	.99929
22	1.00355	1.03049	.99840	.99884
23	.93763	1.04694	.99881	.99834
24	.81414	.86065	.99810	.99843
25	1.07431	1.08961	.99819	.99814
26	1.02921	.99834	.99824	.99790
27	.97914	.99854	.99797	.99823

(Continued)

TABLE 8.4 (Cont.)

28	1.08331	1.06731	.99782	.99811
29	1.07629	1.02820	.99795	.99776
30	.90522	.93443	.99782	.99787
31	.87070	.90980	.99790	.99778
32	1.06103	1.03555	.997936	.99794
33	.99899	.98297	.99815	.99768
34	1.05561	1.10186	.99784	.99751
35	.96562	1.06630	.99763	.99733
36	.88838	.87818	.99728	.99732
37	1.05367	1.12275	.99691	.00742
38	.91351	.94498	.99711	.99699
39	1.05124	1.03427	.99704	.99717
40	1.06043	.99126	.99726	.99675
41	1.02913	.981795	.99668	.99668
42	.90451	.97423	.99689	.99675
43	1.05571	1.06397	.99660	.99703
44	.94088	.86578	.99669	.99689
45	.92513	1.02653	.99672	.99728
46	.88697	.90656	.99682	.99695
47	.98493	1.02502	.99671	.99698
48	.885353	.91209	.99672	.99698
49	.95399	.91775	.99667	.99705
50	1.001805	1.02429	.99689	.99705

$$\theta(t) = \begin{pmatrix} \theta_1(t) \\ \theta_2(t) \end{pmatrix}, \tag{8.75}$$

while the last two columns are the posterior conditional means

$$m(t) = \begin{pmatrix} m_1(t) \\ m_2(t) \end{pmatrix}, \tag{8.76}$$

where the conditional means are given by (8.66).

Referring to Table 8.4, the last two columns are estimates of the first two, thus consider row 5, then it is seen that .99602 is an estimate of .99791 and that .99585 estimates .96601. It is obvious that there is very little variation because of the large precision of $V(t)$, given by the precision matrix p_v (8.74).

We now show how R is used to generate 100 observations from the bivariate linear dynamic model (8.70), (8.71), and (8.72)

RC 8.2

```
mu<-c(0,0)
cov<-matrix(c(1,0,0,1),nr=2)
y<-x<-u<-v<-mvrnorm(100,mu,cov)
F<-matrix(c(3.5,3,3,3),nr=2)
G<-matrix(c(.5,.5,.5,.5),nr=2)
for ( t in 2:100) x[t]<-G*x[t-1]+v[t]
for ( t in 2:100),y[t]<-F*x[t]+u[t]
```
The 100 bivariate observations are:
```
 [1,] -1.24803569 -0.371835811
 [2,] 0.41000933 -0.412565592
 [3,] -0.81668984 1.805989220
 [4,] 3.83633476 1.048460924
 [5,] 1.71393637 2.319903342
 [6,] -0.98925133 -1.320794117
 [7,] -0.54135602 1.082216520
 [8,] -1.78881329 -1.653353876
 [9,] -7.13598235 1.029486418
[10,] -2.93604016 -1.220828440
[11,] -5.67028886 -1.573298975
[12,] -0.89323951 -1.485885190
[13,] -1.69453152 0.488676039
[14,] -2.93610292 -1.940075140
[15,] -1.79399086 0.034597479
[16,] -1.23215011 0.335029230
[17,] -3.89941606 1.378573948
[18,] 5.24366491 -0.099716618
[19,] 1.49602543 0.160827994
[20,] 4.67709113 0.250896774
[21,] 5.28293959 0.703111726
[22,] 3.88526778 -0.036846095
[23,] 0.41636195 -0.309825756
[24,] -1.35331534 -1.010722335
[25,] 1.42794101 1.558980429
[26,] -5.90595263 0.369317151
```

[27,] 1.61391926 0.312833474
[28,] 4.72506189 −0.098481459
[29,] 3.09897964 −0.059348565
[30,] −0.87040536 1.621137626
[31,] −1.35666324 1.086472380
[32,] 1.38726994 −0.402466318
[33,] 3.43796830 0.067533288
[34,] −0.41255020 −0.450530245
[35,] 3.58622086 −0.484720664
[36,] 10.53482204 −0.440642976
[37,] 9.36753418 −0.972734748
[38,] 0.88012504 −0.736901554
[39,] −1.39589930 −0.366509299
[40,] −0.79120801 0.412425631
[41,] −9.61839091 −1.147238511
[42,] −9.39961601 −0.658957975
[43,] −0.28955335 −0.840967380
[44,] 1.40729744 0.843116882
[45,] −0.88145595 0.954135802
[46,] −2.94565548 −1.015134135
[47,] −0.70047295 −0.135172694
[48,] 10.24904394 −0.513429379
[49,] 5.23970932 −1.572817005
[50,] −2.10182681 −0.851122427
[51,] −5.75611717 −0.552913475
[52,] −10.27146934 0.432185744
[53,] −10.06993237 −0.283973018
[54,] 2.58533080 −0.227332768
[55,] −1.05289442 −0.610443279
[56,] 3.43039811 0.644530576
[57,] 0.31942729 −0.071740757
[58,] 4.38275651 −0.084259015
[59,] 5.85604550 −0.002621604
[60,] 7.08614578 0.216908509
[61,] 1.55670973 −1.049302226
[62,] 4.92721551 0.225676994
[63,] 1.93404748 −0.241970096
[64,] 1.45519193 1.339634759
[65,] −8.38030326 0.417468956
[66,] −0.41070962 0.504981648
[67,] 3.10543438 −0.007493484
[68,] −7.37655321 0.268931641
[69,] −4.88627093 −1.089407419
[70,] 1.15897920 0.527885692
[71,] −0.01668236 0.977468411

[72,] 2.29816427 –0.940163039
[73,] 6.07647541 –0.160693416
[74,] 1.80314071 –1.199159821
[75,] –3.64434859 –0.766768490
[76,] 0.95970508 2.959125416
[77,] 2.61055776 0.406540363
[78,] –3.03864371 –0.183757180
[79,] 7.01260326 –0.734379281
[80,] 3.87322000 –1.786121088
[81,] 4.56156229 –0.739215026
[82,] –4.91290890 –0.539164444
[83,] –4.08129690 0.939827151
[84,] 0.71886244 –1.272106121
[85,] 5.73463360 0.651265699
[86,] 5.36466158 –0.972741338
[87,] –0.24511978 1.057173926
[88,] 9.15434440 0.306357930
[89,] 0.23879106 –0.707087530
[90,] 5.10392055 –0.709222035
[91,] 1.48980125 0.219072084
[92,] 1.92650292 1.000252083
[93,] –6.41826991 1.355878087
[94,] –0.89635435 –0.008785834
[95,] 0.08684406 –0.520597987
[96,] –5.15800725 1.807832941
[97,] 6.45243665 1.050226541
[98,] 4.11201916 –0.761849840
[99,] 4.50415709 0.958544153
[100,] 1.32661981 –0.620577777
Now corresponding to this are the 100 system bivariate state vectors:
[1,] –1.248035692 –0.371835811
[2,] –0.047557448 –0.412565592
[3,] –0.186770792 1.805989220
[4,] 0.831766524 1.048460924
[5,] 0.473293252 2.319903342
[6,] –0.167245490 –1.320794117
[7,] –0.138884171 1.082216520
[8,] –0.412945639 –1.653353876
[9,] –1.631656705 1.029486418
[10,] –0.833748558 –1.220828440
[11,] –1.352702919 –1.573298975
[12,] –0.348797994 –1.485885190
[13,] –0.415317893 0.488676039
[14,] –0.698613748 –1.940075140
[15,] –0.476288384 0.034597479

[16,] −0.326732068 0.335029230
[17,] −0.902840466 1.378573948
[18,] 1.064943261 −0.099716618
[19,] 0.450777124 0.160827994
[20,] 1.089439932 0.250896774
[21,] 1.295035456 0.703111726
[22,] 1.007285669 −0.036846095
[23,] 0.204445509 −0.309825756
[24,] −0.278020574 −1.010722335
[25,] 0.286429050 1.558980429
[26,] −1.280608468 0.369317151
[27,] 0.216358894 0.312833474
[28,] 1.074053631 −0.098481459
[29,] 0.808001435 −0.059348565
[30,] −0.103645476 1.621137626
[31,] −0.312996885 1.086472380
[32,] 0.273504777 −0.402466318
[33,] 0.794382375 0.067533288
[34,] −0.003413113 −0.450530245
[35,] 0.796558733 −0.484720664
[36,] 2.429578091 −0.440642976
[37,] 2.351627384 −0.972734748
[38,] 0.456875274 −0.736901554
[39,] −0.259435925 −0.366509299
[40,] −0.204650216 0.412425631
[41,] −2.160159114 −1.147238511
[42,] −2.328821238 −0.658957975
[43,] −0.323103104 −0.840967380
[44,] 0.276832419 0.843116882
[45,] −0.165119942 0.954135802
[46,] −0.672936766 −1.015134135
[47,] −0.230431408 −0.135172694
[48,] 2.251961831 −0.513429379
[49,] 1.414597831 −1.572817005
[50,] −0.309895087 −0.851122427
[51,] −1.313569937 −0.552913475
[52,] −2.428500957 0.432185744
[53,] −2.507596189 −0.283973018
[54,] 0.295896156 −0.227332768
[55,] −0.201099188 −0.610443279
[56,] 0.739966337 0.644530576
[57,] 0.153202324 −0.071740757
[58,] 0.990968371 −0.084259015
[59,] 1.411451040 −0.002621604
[60,] 1.731526955 0.216908509

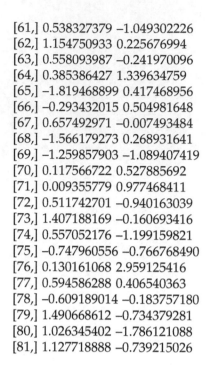

[61,] 0.538327379 −1.049302226
[62,] 1.154750933 0.225676994
[63,] 0.558093987 −0.241970096
[64,] 0.385386427 1.339634759
[65,] −1.819468899 0.417468956
[66,] −0.293432015 0.504981648
[67,] 0.657492971 −0.007493484
[68,] −1.566179273 0.268931641
[69,] −1.259857903 −1.089407419
[70,] 0.117566722 0.527885692
[71,] 0.009355779 0.977468411
[72,] 0.511742701 −0.940163039
[73,] 1.407188169 −0.160693416
[74,] 0.557052176 −1.199159821
[75,] −0.747960556 −0.766768490
[76,] 0.130161068 2.959125416
[77,] 0.594586288 0.406540363
[78,] −0.609189014 −0.183757180
[79,] 1.490668612 −0.734379281
[80,] 1.026345402 −1.786121088
[81,] 1.127718888 −0.739215026

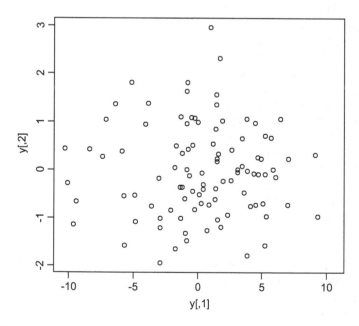

FIGURE 8.3
Bivariate Response for Model (7.70)

```
[82,] -0.966455435 -0.539164444
[83,] -1.014338803 0.939827151
[84,] 0.047042898 -1.272106121
[85,] 1.279590010 0.651265699
[86,] 1.334323686 -0.972741338
[87,] 0.093787125 1.057173926
[88,] 2.044719548 0.306357930
[89,] 0.280255741 -0.707087530
[90,] 1.165344094 -0.709222035
[91,] 0.460549621 0.219072084
[92,] 0.479283941 1.000252083
[93,] -1.373028432 1.355878087
[94,] -0.351748571 -0.008785834
[95,] -0.019784496 -0.520597987
[96,] -1.148422111 1.807832941
[97,] 1.306272354 1.050226541
[98,] 1.058923409 -0.761849840
[99,] 1.118581954 0.958544153
[100,] 0.419091286 -0.620577777
> x
```

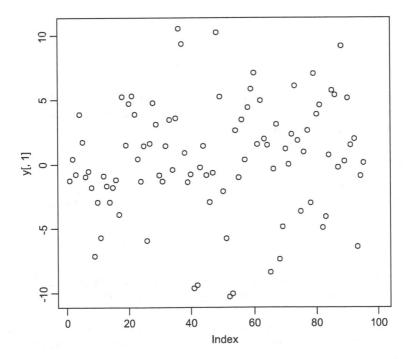

FIGURE 8.4
Response of First Component of the Model (8.70)

The plot of the bivariate y response is portrayed in Figure 8.3.

The student will be asked to compute the covariance and correlation between the first and second components of the response. See Figure 8.4.

Using these bivariate observations and states from the general dynamic model, the student will be asked to perform a Bayesian analysis that estimates the states of the system as well as the parameters of the observation and system errors. See Chapter 11 of West and Harrison[13] on the Bayesian approach to estimating the parameters of a multivariate dynamic linear model.

8.10 Testing Hypotheses

Consider the system equation

$$\theta(t) = \beta\theta(t-1) + v(t), t = 1, 2, \ldots \tag{8.77}$$

where β is an unknown parameter and $|\beta| < 1$. Recall that the restriction $|\beta| < 1$ ensures the process is stationary. Also, assume that the $v(t) \sim nid(0, \tau)$, where $t=1,2, \ldots$ and $\tau > 0$ is the known precision parameter.

We are interested in testing the null hypothesis

$$H_0 : \beta = .5$$

Versus $\tag{8.78}$

$$H_1 : \beta \neq .5.$$

Use the information generated by RC 8.1 where $\beta = .5$ was the assumed value for the autoregressive parameter of the system equation. The code and the 25 states are given below:

RC 8.3.

```
> x<-v<-rnorm(25)
```
Note that the vector x includes the 25 states of the system equation (8.77).
```
> x<-v<-rnorm(25)
> for ( t in 2:25) { x[t]<-0.5*x[t-1]+v[t]
+ }
> x
```
 [1] 0.31333278 2.75463086 0.82144054 1.34697160 −0.72060105 −1.05783691

 [7] −0.04589585 −1.16427438 0.69077320 −0.57384234 −1.92647815 − 0.27850692

[13] −0.93883024 −0.52529289 1.80098032 0.35961288 0.94063136 −1.14001415

[19] −0.48476299 −1.73904220 −0.15354250 −0.85198102 −0.64718522 − 2.56236746

[25] –1.70908470
> sum(x)
[1] –7.491165
The following generates the square of each state
> y<-rnorm(25)
> for (t in 2:25) {y[t]<-x[t]*x[t]}
> y
  [1]   0.736792788   7.587991171   0.674764566   1.814332485   0.519265873
1.119018936
  [7]   0.002106429   1.355534825   0.477167615   0.329295028   3.711318057
0.077566106
[13]   0.881402221   0.275932616   3.243530129   0.129321425   0.884787352
1.299632260
[19]   0.234995157   3.024267777   0.023575299   0.725871653   0.418848713
6.565727023
[25] 2.920970502
The following generates the product of each state and its predecessor
> z<-rnorm(25)
> for (t in 2:25) {z[t]<-x[t]*x[t-1]}
> z
 [1] 4.37930302 0.86311616 2.26276547 1.10645708 –0.97062915 0.76227839
 [7] 0.04855032 0.05343536 –0.80424954 –0.39639491 1.10549472 0.53653750
[13] 0.26147072 0.49316085 –0.94604215 0.64765573 0.33826315 –1.07233306
[19] 0.55263667 0.84302330 0.26701689 0.13081530 0.55138952 1.65832636
[25] 4.37930302
The following generates the square is each preceding state.
> w<-rnorm(25)
> for (t in 2:25) w[t]<-x[t-1]*x[t-1]
> w
 [1] –0.702546858 0.098177433 7.587991171 0.674764566 1.814332485
 [6] 0.519265873 1.119018936 0.002106429 1.355534825 0.477167615
[11] 0.329295028 3.711318057 0.077566106 0.881402221 0.275932616
[16] 3.243530129 0.129321425 0.884787352 1.299632260 0.234995157
[21] 3.024267777 0.023575299 0.725871653 0.418848713 6.565727023
The following generates the sum of the state values
> sum(x)
[1] –7.491165
The following is the sum of the squares of the state values
> sum(y)
[1] 39.03402
The following is the sum of the product of each state and its proceeding
value
> sum(z)
[1] 17.05135
The following is the sum of the squares of the proceeding states

\> sum(w)
[1] 34.77188

Remember that the data is the 25 values of the states of the system. In terms of the state values, one may show that

$$\sum_{t=2}^{t=25} \theta(t)\theta(t-1) = 17.0513 \tag{8.79}$$

$$\sum_{t=2}^{t=25} \theta^2(t-1) = 34.7718, \tag{8.80}$$

and

$$\sum_{t=2}^{t=25} \theta^2(t) = 39.0340. \tag{8.81}$$

Also, it can be shown that the likelihood function is given by

$$g(\theta|\beta) = (\tau/2\pi)^{n/2} \exp -(\tau/2) \sum_{t=2}^{t=25} \theta^2(t-1)[(\beta-h)^2 + k] \tag{8.82}$$

where $\theta = [\theta(1), \theta(2), \ldots, \theta(25)]'$,

$$h = [\sum_{t=2}^{t=25} \theta(t)\theta(t-1)]/\sum_{t=2}^{t=25} \theta^2(t-1) = 17.0513/34.7718 = .490377, \tag{8.83}$$

and

$$k = \sum_{t=2}^{t=25} \theta^2(t) - [\sum_{t=2}^{t=25} \theta(t)\theta(t-1)]^2 / \sum_{t=2}^{t=25} \theta^2(t-1) = \tag{8.84}$$
$$39.0340 - 17.0513 * 17.0351/34.7718 = 30.68832.$$

The above information is needed for testing the hypotheses stated by (8.78), and the approach explained by Lee[21, ch. 4] (a formal way to test hypotheses from a Bayesian approach) will be closely followed.

Let π_0 be the prior probability of the null hypothesis and that of the alternative be $\pi_1 = 1 - \pi_0$, and denote the posterior probability of the null as

$$p_0 = [\pi_0 g(\theta|\beta_0)]/[\pi_0 g(\theta|\beta_0) + \pi_1 g_1(\theta)], \tag{8.85}$$

where $g(\theta|\beta)$ is given by (8.82) and $g(\theta|\beta_0) = g(\theta|\beta = .5)$.

In addition, let

$$g_1(\theta) = \int_{-1}^{+1} p_1(\beta)g(\theta|\beta)d\beta \qquad (8.86)$$

where $p_1(\beta)$ is the prior density of $p_1(\beta)$ under the alternative hypothesis. Also,

$$g(\theta|\beta) = (\tau/2\pi)^{n/2}e^{-(\tau/2)34.7718[(\beta-.49037)^2+30.6882]}, \qquad (8.87)$$

and

$$g(\theta|\beta = .5)] = (\tau/2\pi)^{n/2}e^{-(\tau/2)1067.105}. \qquad (8.88)$$

In addition, it can be shown that the predictive density (8.86) reduces to

$$g_1(\theta) - [1/\sum_{t=2}^{t=n} \theta^2(t-1)]^{n/2}e^{-\tau k/2} = [1/34.7718]^{n/2}e^{-\tau 15.3441}. \qquad (8.89)$$

Remember that τ and n are known, and that there is sufficient information to calculate the posterior probability p_0(8.85) of the null hypothesis, namely

$$p_0 = [1/1 + (\pi_1 g_1(\theta)/\pi_0 g(\theta|\beta = .5)]. \qquad (8.90)$$

Assuming the prior probability of the null hypothesis is ½, it can be shown that

$$p_0 = [1/1 + (\pi_1 g_1(\theta)/\pi_0 g(\theta|\beta = .5)], \qquad (8.91)$$

where

$$g_1(\theta) = 2.38026640869944e^{-26} \qquad (8.92)$$

and

$$g(\theta|\beta = .5)] = 1.16882542299e^{-238}. \qquad (8.93)$$

What is the posterior probability of the null hypothesis? The student is referred to problem 14 in the exercises to explain the answer to the above question.

8.11 Summary

The chapter begins with a brief introduction, then the next section defines the dynamic linear model, followed by an explanation of estimation, filtering, smoothing, prediction, the control problem, and the Kalman filter. Next, R is employed to generate observations and states from a univariate dynamic linear model; then using those observations, a Bayesian analysis is performed that results in estimating the states, and Table 8.3 reports the posterior analysis. This is followed by a definition of the control problem, where the system equation is generalized to include the control variable. The control variable is chosen so that the mean of the posterior distribution of a state is equal to the preset target value. Figure 8.1 portrays the Kalman filter estimates of the states defined by the dynamic linear model (8.55). The last part of the chapter introduces the subject to adaptive estimation. This is a generalization of the model where the parameters of the variances of the observation and system errors are unknown. The Bayesian approach is to find the posterior distribution of the states as well as the precision of the observation and system errors. The chapter concludes with an example of adaptive estimation, where the observations and states are bivariate. Then, RC 8.2 generates 50 bivariate observations and states with known F and G matrices and known 2×2 precision matrices for the observation and system equations. Testing hypotheses is an integral part of Bayesian inference and is illustrated with a univariate dynamic linear model, where the null and alternative hypotheses involve the design matrix of the system equation.

8.12 Exercises

1. In two pages, describe the content of Chapter 8.
2. Equations (8.1) and (8.20) define the dynamic linear model:
    (a) Why is this model called linear?
    (b) What is the role of F in the observations equation?
    (c) What is the role of G in the observation equation?
3. Describe in detail the following.
    (a) Filtering.
    (b) Smoothing.
    (c) Prediction
    (d) The control problem.
    (e) The Kalman filter.
    (f) Explain the recursive nature of the Kalman filter.
    (g) Given the previous state, what is the posterior distribution of the present state?
4. Refer to Table 8.1 and explain how to calculate the entries of the columns U, V, θ, m, and P.

5. Refer to the R code R 8.1, and generate 100 observations generated by the R statements for Dynamic linear model (8.52) and plot the 100 observations over time from 1 to 100.

6. Refer to the WinBUGS code BC 8.1. Using 45,000 observations for the simulation with 5,000 initial values, perform the Bayesian analysis as indicated. Use the 100 observations generated with RC 8.1 as data and note the prior distributions for the parameters as given in the program statements of BC 8.1. Your results of the posterior analysis should be similar to those reported in Table 8.3.

7. Refer to Table 8.3.

   (a) What is the posterior mean and 95% credible interval for β, the design matrix of the observation equation?

   (b) What is the posterior distribution of delta, the design matrix for the system equation.

   (c) The precision of the errors for the observation equation is tau1. What is the posterior median and 95% credible interval for this parameter?

   (d) The 'true' value (that used to generated the data) is 1. How well does the Bayesian posterior mean and median estimate tau1.

8. Refer to Table 8.3, the control problem with target value $T = -.7$ for all states. The model is in (8.53).

   (a) Explain how the values of the control variable are chosen at each state.

   (b) Describe the six steps for choosing the value of the control variable at each time.

   (c) At time 5, what value is chosen for the control variable?

9. With regard to the linear dynamic model defined by (8.54), explain what is meant by adaptive estimation.

   (a) Given $\theta(0)$, the mean $m(0)$, and the precision $p(0)$ at the initial state, derive the posterior distribution of $\theta(1)$. See (8.55) and (8.56) for the mean and precision respectively.

   (b) Derive the posterior distribution of the first k states $\theta(1), \ldots, \theta(k)$ given by (8.51).

10. Refer to Table 8.4, which is based on the bivariate linear dynamic model (8.70). Fifty observations were generated from (8.70) using the precision matrix $p(u)$ for the errors of the observation equation (8.73) and $p(v)$ the precision matrix for the system errors (8.74).

   a. Describe how the columns of Table 8.4 are computed.

   b. Refer to line 8 of Table 8.4, what value does .9964 estimate?

   c. What value does .9968 estimate.

11. The R code RC 8.2 generates 100 observations from the bivariate dynamic linear model defined by (8.70), (8.71), (8.72), (8.73), and (8.74).

   (a) Your results should be similar to those appearing in Table 8.4.

   (b) Using R, plot the 100 observation for the 100 observations over the 100 times.

12. Using the 100 observations generated by RC 8.2 as data for a Bayesian analysis, execute the posterior analysis with 45,000 observations for the simulation with 5,000 starting values. This generalizes the univariate dynamic linear model to the bivariate case both for the observation and system equations. Refer to the WinBUGS code BC 8.1 which is appropriate for the univariate model.

13. Refer to Section 8.10 and derive the posterior probability p_0 of the null hypothesis.

14. Refer to Equations (8.91), (8.92), and (8.93) and calculate the posterior probability of the null hypothesis (8.91)

References

1. Broemeling, L.D. (1985). *Bayesian Analysis of Linear Models*. Marcel-Dekker Inc., New York.
2. Gajic, Z. (2003). *Linear Dynamic Systems and Signals*. Prentice Hall, New York.
3. Crassidis, J.L. and Junkms, J.L. (2001). *Optimal Estimation of Dynamic Systems*. CRC Press, Chapman & Hall, Boca Raton, FL.
4. Fuchs, A. (2013). *Nonlinear Dynamics in Complex Systems*. Springer-Verlag, Berlin and Heidelberg.
5. Yedavalli, R.K. (2014). *Robust Control of Uncertain Dynamic Systems. A Linear State Approach*. Springer-Verlag, New York, London.
6. Aoki, M. (1967). *Optimization in Stochastic Systems, Topics in Discrete Time Series*. Academic Press, New York, London.
7. Jazwinski, A.H. (1979). *Stochastic Processes and Filtering Theory*. Academic Press, New York.
8. Harrison, P.J. and Stevens, C.F. (1976). "Bayesian forecasting" (with discussion), *Journal of the Royal Statistical Society: Series B*, 38, 205–247.
9. Kalman, R.E. (1960). "A new approach to linear filtering and prediction", *Transactions of the ASME – Journal of Basic Engineering*, 82(D), 33–45.
10. Petris, G., Petrone, S. and Campagnoli, P. (2009). *Dynamic Linear Models with R*. Springer Verlag, Berlin.
11. Thiele, T. (1880). "Om anvendelse af mindset kradraters methode I nogle tilflde, hvor en komplikation af vise slags uensartede tilfldige fejlkilder giver fejleneen systematisk karakter", in *Det Kongelige Danske Videnskabernes Selskabs Skkrifter Naturvidenskabelig Og Mathematisk Afdeling*, 381–408, English translation in *Thiele, Pioneer in Statistics*, edited byS.L.Lauritzen, University Press, Oxford, UK, 2001.

12. Cowpertwait, P.S.P. and Metcalfe, A.V. (2009). *Introductory Time Series with R.* Springer, New York.
13. West, M. and Harrison, J. (1997). *Bayesian Forecasting and Dynamic Models, Second Edition.* Springer, New York.
14. Maybeck, P.S. (1979). *Stochastic Models, Estimation, and Control. Volumes I and II.* Academic Press, New York.
15. Steigerwald, D.G. (1992). "Adaptive estimation in time series regression models", *Journal of Econometrics*, 54, 251–275.
16. Alspach, D.I. (1974). "A parallel filtering algorithm for linear systems with time-varying noise statistiscs", *IEEE Transactions on Automatic Control*, AC-19, 555–556.
17. Hawkwa, R.M. (1973). *Demodulation of Pulse Modulated Signals Using Adaptive Estimation.* Unpublished master's thesis, University of Newcastle.
18. Dreze, J.H. (1977). "Bayesian regression analysis using poly-t densities", in *New Developments in the Application of Bayesian Methods*, edited byA.Aykac and C. Bremat, North Holland Publishing Company, Amsterdam.
19. Box, G.E.P. (1973). *Bayesian Inference in Statistical Analysis.* Addison-Wesley Publishing Company, Menlow Park, CA.
20. Zellner, A. (1971). *An Introduction to Bayesian Inference in Econometrics.* John Wiley & Sons Inc., New York.
21. Lee, P.M. (1997). *Bayesian Statistics, An Introduction, Second Edition.* John Wiley & Sons Inc., New York.

9

The Shift Point Problem in Time Series

9.1 Introduction

This chapter introduces the reader to the so-called shift point problem in time series. The shift point problem is that of determining the time point where the parameters of the time series model change and this point is referred to as the shift point. This is a vast area of interest in time series and has been studied from the mid-1950s. This chapter is focused on a shift in parameters over time, while Chapter 10 deals with changes that depend on the actual value of the time series. This area is called threshold changes, where the parameters of the series change when the value of the series exceeds a given value called the threshold.

Page[1],[2] first discussed the problem in the context of quality control, and statistical work in this area had advanced rapidly as demonstrated by Page who developed non-parametric procedures for detecting shifts for univariate sequences. After that, Chernoff and Zacks[3] and then Kander and Zacks[4] developed Bayesian methods to the test the current mean and test for a shift in the mean ($H_0 : \mu_1 = \mu_2$ versus $H_1 : \mu_1 \neq \mu_2$) of a sequence of independent random variables.

Of course, our main emphasis is on the Bayesian approach to time series; thus, consider the following citations: Chin Choy and Broemeling,[5] Booth and Smith,[6] Broemeling,[7],[8],[9] Shao and Zhang,[10] and Barberi and Conigliani.[11]

The chapter begins with a shifting of normal sequence with one shift point followed by a series that follows an autoregressive process of order one with one change point. The moving average process MA(1) is considered next, where the posterior distribution of all parameters is derived under the assumption that the shift point M has a uniform distribution over the range $1, 2, \ldots, n-1$, where n is the number of observations for the time series. The posterior mass function of the shift point is derived as is the posterior density of the autoregressive parameter, which is shown to be a mixture of t densities, where the mixing distribution is that of the posterior mass function of the shift point. In a similar manner, the Bayesian analysis for the MA(1) series is presented, where the posterior distribution of the shift point and moving average parameter is derived. It is assumed that the shift point has a discrete prior uniform distribution and that for the moving average parameter is informative. In each case, the posterior analysis is illustrated by generating observations from the model using R where the parameters of the model are known, then followed by a Bayesian analysis with WinBUGS. The next section focuses on testing hypotheses about the parameters of the shifting series. For example, for the AR(1), there is enough evidence to conclude that two regimes are actually the same.

Next to be considered are regression time series models with autoregressive and moving average errors, followed by a regression model with seasonal effects where the residuals follow either an AR(1) series or a MA (1) process. In both cases, there is one shift. The last section of Chapter 9 presents many examples in economics taken from the book by Broemeling and Tsurumi.[12] The chapter concludes with a thorough explanation of the threshold autoregressive model, which shifts parameters according to the level of the present or related observation values, and many exercises that complement and expand the subject matter presented earlier.

9.2 A Shifting Normal Sequence

First to be considered is a simple example of structural change in a sequence of n.i.d random variables, where the first $m Y(1), Y(2), ..., Y(m)$ follow a $N(\theta_1, \tau)$, while the remaining $Y(m+1), Y(m+2), ..., Y(n)$ follow a $N(\theta_2, \tau)$, and $n > 3$. The parameter of primary interest is the shift point m, and the pre-change and post-change means.

It is easy to see that the likelihood function is

$$L(\theta_1, \theta_2, \tau | s) \propto \tau^{n/2} \exp -(\tau/2)\{\sum_{t=1}^{t=m} [y(t) - \theta_1]^2 + \sum_{t=m+1}^{t=n} [y(t) - \theta_2]^2\} \quad (9.1)$$

where s is the sample, θ_1 and $\theta_2 \in R$, $\tau > 0$, and $m = 1, 2, ..., n-1$. The Bayesian analysis involves deriving the posterior mass function of m and the marginal posterior distributions of θ_1, θ_2, and τ.

Consider the following prior distribution for $\theta = (\theta_1, \theta_2), \tau$, and m: the prior for $\theta = (\theta_1, \theta_2)$ and τ is normal-gamma, that is

$$g(\theta | \tau) \propto \tau^{n/2} \exp -(\tau/2)(\theta - \mu)' P(\theta - \mu), \theta \in R^2 \quad (9.2)$$

$$g(\tau) \propto \tau^{\alpha-1} \exp -\tau\beta, \tau > 0, \quad (9.3)$$

and

$$g(m) = (n-1)^1, m = 1, 2, \ldots, n-1, n > 3 \tag{9.4}$$

Note that the hyper-parameters are the 2×2 positive definite precision matrix P, the 2×1 real mean vector μ and the parameters of the gamma prior, namely α and β.

Combining the likelihood with the prior distribution results in the joint posterior density

$$g(\theta, \tau, m|s) \propto \tau^{(n+2+2\alpha)/2-1} \exp -(\tau/2)\{[\theta - A^{-1}(m)B(m)]' \\ A(m)[\theta - A^{-1}(m)B(m)] + C(m) - B'(m)A^{-1}(m)B(m)\} \tag{9.5}$$

where s is the vector of sample values,

$$A(m) = P + \begin{pmatrix} m, 0 \\ 0, n-m \end{pmatrix} \tag{9.6}$$

$$B(m) = \left[\sum_{t=1}^{t=m} y(t) \sum_{t=m+1}^{n} y(t) \right]. \tag{9.7}$$

and

$$C(m) = 2\beta + \sum_{t=1}^{t=n} y^2(t) + \mu'P\mu. \tag{9.8}$$

By integrating the joint posterior density of the parameters with respect to θ and τ, the marginal posterior mass function of m is

$$g(m|s) \propto 1/|A(m)|^{1/2}\{C(m) - B'(m)A^{-1}(m)B(m)\}^{(n+2\alpha)/2} \tag{9.9}$$

for $m=1,2,\ldots,n-1$.

In addition, the conditional posterior density of θ is

$$f(\theta|m,s) \propto \{[\theta - A^{-1}(m)B(m)]'A(m)[\theta - A^{-1}(m)B(m)] + \\ C(m) - B'(m)A^{-1}(m)B(m)\}^{-(n+2+2\alpha)/2} \tag{9.10}$$

This is a bivariate t density with $n+2\alpha$ degrees of freedom, location parameter $A^{-1}(m)B(m)$, and precision matrix

$$T(m) = (n + 2\alpha)A(m)/[C(m) - B'(m)A^{-1}(m)B(m)] \qquad (9.11)$$

One may show that the conditional distribution of τ is gamma with parameters $(n + 2\alpha)/2$ and $[C(m) - B'(m)A^{-1}(m)B(m)]/2$.

In order to determine the marginal distributions of θ and τ, one may average the conditional distributions of these parameters with respect to the marginal distribution of the shift point; thus, for θ, the density is

$$g(\theta|s) = \sum_{m=1}^{m=n-1} g(m|s)f(\theta|m,s), \qquad (9.12)$$

where $f(\theta|m,s)$ is given by (9.10). Thus, the marginal posterior density of θ is a mixture of $n-1$ t-distributions with mixing distribution which is the marginal posterior mass function of m.

As for the precision parameter, its posterior marginal density is

$$g(\tau|s) = \sum_{m=1}^{m=n-1} g(m|s)f(\tau|m,s), \qquad (9.13)$$

where $f(\tau|m,s)$ is a gamma density with parameters $\alpha(m) = (n + 2\alpha)/2$ and $\beta(m) = [C(m) - B'(m)A^{-1}(m)B(m)]/2$.

Although the case of a shifting sequence of normal random variables is not realistic for a time series, since for a time series one would expect to have a non-zero autocorrelation function, thus the next situation is more realistic and to be considered is a shifting autoregressive time series.

9.3 Structural Change in an Autoregressive Time Series

Recall from Chapter 4 the Bayesian posterior analysis for the autoregressive AR(p) time series. Consider the following AR(1) time series with one shift point defined by

$$Y(t) = \theta_{11} + \theta_{12}Y(t-1) + e(t), t = 1, 2, ..., m$$
$$Y(t) = \theta_{21} + \theta_{22}Y(t-1) + e(t), t = m+1, ..., n \qquad (9.14)$$

the shift point is m, where $m=1,2, ..., n-1$. The autoregressive parameters are for the first phase

$\theta_1 = \begin{pmatrix} \theta_{11} \\ \theta_{12} \end{pmatrix}$, and for the second phase $\theta_2 = \begin{pmatrix} \theta_{21} \\ \theta_{22} \end{pmatrix}$, while the errors are

the $e(t) \sim nid(0, \tau), \tau > 0, t = 1, 2, ..., n$.

$y(0)$ is the initial value and all the other parameters θ_1, θ_2, and τ are considered unknown as well as is the unknown shift point m.

Thus, for the first m observations, the time series follows an AR(1) process with parameters θ_1, and for the remaining $n-m$ observations, the series follows a distinct AR(1) process with parameters θ_2. A priori the following distributions are assigned to the parameters: Suppose that the shift point m and the vector (θ, τ) are independent, where $\theta = \begin{pmatrix} \theta_1 \\ \theta_2 \end{pmatrix} \in R^4$, and that m has a uniform distribution over the integers $1, 2, \ldots, n-1$. In addition, assume that the four-variate vector θ with mean

$$\mu = \begin{pmatrix} \mu_{11} \\ \mu_{12} \\ \mu_{21} \\ \mu_{22} \end{pmatrix} \tag{9.15}$$

and the 4×4 precision matrix

$$P = \begin{pmatrix} p_{11}, p_{12} \\ p_{21}, p_{22} \end{pmatrix} \tag{9.16}$$

where p_{11} is 2×2. Also, finally, assign a gamma distribution to τ, namely $\tau \sim gamma(\alpha, \beta)$. Note the hyper-parameters μ, P, and (α, β) are known.

It can be shown that the marginal posterior mass function of the shift point is

$$g(m|s) \propto C_1^{-(2a+n)/2}(m)|A(m)|^{-1/2}, m = 1, 2, \ldots, n-1 \tag{9.17}$$

where

$$A(m) = \begin{pmatrix} A_{11}, P_{12} \\ P_{21}, A_{22} \end{pmatrix},$$

$$A_{11} = P_{11} + G_1,$$

$$A_{22} = P_{22} + H_1,$$

$$G_1 = \begin{bmatrix} m, \sum_{t=1}^{t=m} Y(t-1) \\ \sum_{t=1}^{t=m} Y(t-1), \sum_{t=1}^{t=m} Y^2(t-1) \end{bmatrix},$$

$$H_1 = \begin{bmatrix} n - m, \ \sum\limits_{t=m+1}^{t=n} Y(t-1) \\ \sum\limits_{t=m+1}^{t=n} Y(t-1), \ \sum\limits_{t=m+1}^{t=n} Y^2(t-1) \end{bmatrix},$$

$$C_1(m) = b + (1/2)[S_1(\tilde{\theta}_1) + S_1(\tilde{\theta}_2) + \mu'P\mu + \tilde{\theta}_1' \, G_1\tilde{\theta}_1 + \tilde{\theta}_2' \, H_1\tilde{\theta}_2] - \tilde{\theta}' A(m)\theta' \tag{9.18}$$

$$S_1(\tilde{\theta}_1) = \sum_{t=1}^{t=m} [Y(t) - \tilde{\theta}_{11} - \tilde{\theta}_{12}Y(t-1)]^2,$$

$$S_1(\tilde{\theta}_2) = \sum_{t=m+1}^{t=n} [Y(t) - \tilde{\theta}_{21} - \tilde{\theta}_{22}Y(t-1)]^2,$$

$$\tilde{\theta}_1 = G_1^{-1}[\sum_{t=1}^{t=m} Y(t), \sum_{t=1}^{t=m} Y(t)Y(t-1)],$$

and

$$\tilde{\theta}_2 = H_1^{-1}[\sum_{t=m+1}^{t=n} Y(t), \sum_{t=m+1}^{n} Y(t)Y(t-1)]. \tag{9.19}$$

The posterior density of θ is easily determined as a mixture of multivariate t densities and in a similar manner it can be stated that the marginal density of τ is the mixture

$$g(\tau|s) = \sum_{m=1}^{m=n-1} g(\tau|m,s)g(m|s), \tau > 0$$

where the marginal posterior mass function of the shift point is (9.17).

To illustrate the posterior distribution of the shift point, consider the following model:

$$Y(t) = 4 + \theta_1 Y(t-1) + \varepsilon(t), t = 2,3,...,10$$
$$Y(t) = 4 + \theta_2 Y(t-1) + \varepsilon(t), t = 11,...,20 \tag{9.20}$$

where $\varepsilon(t) \sim n.i.d.(0,1)$, and $\theta_1 \neq \theta_2$.

Now let $\theta_1 = 0.5$, $\theta_2 = -0.5$, and $\tau = 1$, then the following R code generates data from this model:

RC 9.1.

For the first regime:
y1<-e1<-rnorm(25)
for (t in 2:10) y1[t]<-4+0.5*y1[t-1]+e1[t]
The 25 observations are elements of the vector y1:
    y1=(−0.14793803, 2.82127477, 5.58927600, 5.97686344, 8.28542519, 8.76201910, 8.72427860, 6.60587317, 6.59145071, 6.96931828, 0.46257345, −1.72267551, −1.26713331, −0.63879404, −1.07779636, 1.67229292, 1.04294864, 0.55626107, 0.15911277, 1.54227407, 1.51881178, −0.10349231, −1.33494483, 0.73962146, −0.62969202)
and for the second:
y2<-e2<-rnorm(25)
for (t in 2:10) y2[t]<-4−0.5*y2[t-1]+e2[t]
    y2=(−0.7432287, 5.3450388, 2.2723180, 3.4807999, 2.6765519, 1.4394705, 4.3588370, 1.5804615, 4.5265112, 0.7541239, −0.9162551, −1.5092330, 1.7320730, −0.9591964,   −0.7473424,   −0.4495857,   −1.9115178,   1.7644326,   1.1234451, 0.3391542, −1.1477970, 1.9987876, 0.4264450, −1.1203433,−0.1628587).

Note the first regime is an AR(1) series with autocorrelation .5 while the second is also an AR(1) series but with autocorrelation −.5.

Using the above R generated values, the following BUGS code is executed with 45,000 observations for the simulation with 5,000 initial values.

BC 9.1

```
model{
# AR(1) model
mu[1]<-0
delta<-1
y[1] ~ dnorm(mu[1],tau)
for(t in 2:10) {
mu[t] <- 4+theta1*y[t-1]
y[t] ~ dnorm(mu[t],tau)}
for ( t in 11:20) {mu[t]<-4+theta2*y[t-1]
y[t]~dnorm(mu[t],tau)}

# Prior distribution

theta1 ~ dnorm(0.0, .001)
theta2~dnorm(0.0,.001)
tau ~ dgamma(0.01, 0.01)
}
```

list(y=c(−0.9246405, 1.0517816, 4.4831206, 4.8851545, 7.4150595, 7.8909079, 7.6856879, 7.1174456, 7.3633652, 6.99225, 1.082496, 4.512170, 2.964401, 3.198568, 3.086966, 2.099557, 4.479996, 3.917691, 3.444547, 1.726513))

TABLE 9.1

Posterior Analysis for AR(1) Model.

Parameter	Value	Mean	SD	Error	2½	Median	97½
τ	1	1.014	.3393	.00173	.4643	.9766	1.787
θ_1	0.5	.4636	.0587	.000299	.3468	.4637	.5801
θ_2	−0.5	−0.299	.0858	.0000001	−.47	−.2992	−.1304

Table 9.1 reveals that the posterior median gives fairly accurate estimates of the parameters. Note the second column refers to the values of the parameters used to generate the data of RC 9.1. Take the posterior median of .4637 of the posterior distribution of the first regression coefficient which is quite close to the 'true' value of .5 for the autocorrelation of the first regime of the AR(1) series.

Our challenge now is to employ a random shift point m and estimate its posterior mass function based on the above data generated by RC 9.1.

BC 9.2

```
model
{

for ( j in 1:2){theta[j]~dnorm(0,.001)}

for(i in 2:50) {
J[i]<−1+step(i-k-10)
y[i] ~ dnorm(mu[i],tau)
mu[i] <- 4+theta[J[i]]*y[i-1]}

# Prior distribution
for( i in 1:50){
punit[i]<−1/50}

k~ dcat(punit[])

tau ~ dgamma(0.01, 0.01)
}
```

list(y=c(−0.14793803, 2.82127477, 5.58927600, 5.97686344, 8.28542519, 8.76201910, 8.72427860, 6.60587317, 6.59145071, 6.96931828, 0.46257345, −1.72267551,−1.26713331, −0.63879404, −1.07779636, 1.67229292, 1.04294864, 0.55626107, 0.15911277, 1.54227407, 1.51881178, −0.10349231, −1.33494483, 0.73962146, −0.62969202, −0.7432287, 5.3450388, 2.2723180, 3.4807999, 2.6765519, 1.4394705, 4.3588370, 1.5804615, 4.5265112, 0.7541239, −0.9162551, −1.5092330, 1.7320730, −0.9591964, −0.7473424, − 0.4495857, −1.9115178,

1.7644326, 1.1234451, 0.3391542, −1.1477970, 1.9987876, 0.4264450, −1.1203433, −0.1628587))

list(theta=c(1,2),k=25)

The posterior analysis shown in Table 9.2 reveals that the posterior mean for the shift point k is fairly close to the true value of 25; however, for the precision τ, the 95% credible interval does not include the true value of 1. Actually, the posterior median for θ_2 is quite close to the true value of −0.5. Overall, one could conclude that the Bayesian analysis gives only from fair to poor estimates of the true values of the parameters. Of course, better estimates are possible if the sample size is increased from 25 for each regime, to say 100. This will be demonstrated with exercises at the end of the chapter. Another way to obtain better estimates is to put a nonuniform prior mass function with a 'large' probability corresponding to the 'true' shift point at 25.

Again, consider the model (9.14) with one change over 20 observations where the change occurs at time $m=10$. Take the first 10 observations from the data of BC 9.2 for the first regime, and last 10 observation for the second phase. Now assume the following for the prior distributions of the parameters. For the parameters $\theta = \begin{pmatrix} \theta_1 \\ \theta_2 \end{pmatrix}$ and τ, suppose the conditional distribution of θ given τ is normal with mean vector

$$E(\theta) = \begin{pmatrix} 4 \\ 4 \end{pmatrix}$$

and precision matrix τI_2, and suppose the marginal prior mass function of the shift point m is uniform over the set $\{1, 2, \ldots, 19\}$.

Table 9.3 gives the posterior mass function of the shift point for data generated when $\theta_1 = .5$, $\theta_2 = \theta_1 + \delta$, $\delta = .5, 1, 1.49$, and $\tau = 1$.

Note the effect on the mass function of the shift point as the shift δ varies. One might erroneously conclude the shift point is at 17 when $\delta = .5$ but, when $\delta = 1.49$, the posterior probability at the shift point is .647. Thus, the effect of the shift on the posterior mass function of the shift point behaves as one would expect. Note that for the first regime, the series is stationary, but for the second regime the series corresponds to a non-stationary AR(1) series.

TABLE 9.2

Posterior Analysis for Shift Point of an AR(1) Series

Parameter	Value	Mean	SD	Error	2½	Median	97½
K	25	37.12	15.2	.3301	1	44	50
τ	1	.07516	.01570	.000092	.04789	.07389	.1095
θ_1	0.5	.2312	.1793	.0019	−.1063	.2243	.6056
θ_2	−0.5	−.3821	26.92	.1198	−58.65	−.4012	56.96

TABLE 9.3

Posterior Mass Function of k When $\theta_1 = .5$

m	$\delta = .5$	$\delta = 1$	$\delta = 1.49$
1	.000	.000	.000
2	.016	.011	.002
3	.018	.014	.004
4	.019	.106	.004
5	.015	.013	.003
6	.019	.021	.007
7	.016	.018	.006
8	.024	.043	.023
9	.021	.036	.015
10	.058	.291	.647
11	.082	.347	.278
12	.062	.093	.006
13	.044	.024	.001
14	.037	.014	.000
15	.030	.012	.000
16	.065	.014	.000
17	.469	.029	.001
18	.000	.000	.000
19	.000	.000	.000
20	.000	.000	.000

Source: Broemeling.[9, p. 366]

9.4 One Shift in a MA(1) Time Series

We consider a moving process of order one, defined by

$$X(t) = W(t) + \beta_1 W(t-1), t = 1, 2, \ldots, k$$
$$X(t) = W(t) + \beta_2 W(t-1), t = k+1, \ldots, n \qquad (9.21)$$

where the $W(t)$ are distributed as n.i.d$(0, \tau)$ random variables, β_1 and β_2 are real unknown parameters, and k is the time point after which the series changes to another moving average process. Also, the shift point k is unknown with $k=1,2,\ldots,n-1$. The objective of the Bayesian analysis is to estimate the unknown parameters based on the data below which is generated assuming

$$\beta_1 = 0.8, \beta_2 = -0.8, \tau = 1.$$

RC 9.2

```
> set.seed(1)
> b<-c(.8)
> x<-w<-rnorm(20)
> for ( t in 2:20){
+ for ( j in 1:1) x[t]<-w[t]+b[j]*w[t-j]}
>
> x=(  1.26747046  1.59926817,  -0.30936599,  -2.71169235,  -0.64682899,
0.85501113,  -0.05213715,  0.93088400,  1.57629016,  1.25087828,  1.39409843,
1.51731820,  0.70027402,  -1.92969971,  -0.97165561,  0.43973186,  -0.20069850,
-1.59538879, -1.65475196, 0.03542152),
 set.seed(1)
 b<-c(-.8)
 x<-w<-rnorm(20)
 for ( t in 2:20){
+ for ( j in 1:1) x[t]<-w[t]+b[j]*w[t-j]}
>x
(-0.62645381, 0.68480637, -0.98254327, 2.26378369, -0.94671687, -1.08407460,
1.14380376,  0.3483814 -0.93311517,  -1.717707426,  -0.01487841,  -0.76601347,
1.75609188,  -0.81958170,,  2.89669083,  -0.94487834,  0.01975662,  0.95678842,
0.06615223,  -0.06307563,  0.44385631,  0.04695440,  -0.55114406,  -2.04900368,
2.21130710, -0.55198934, -0.11089252, -1.34611598, 0.69845185, 0.80046160)
```

Based on the first twenty observations selected from each regime, the Bayesian analysis is executed with 45,000 observations for the simulation and 5,000 initial values.

Note the prior distribution for the shift point given as the vector p unit in the code below. The prior distributions for the moving average parameters are somewhat informative with a prior precision of 1. While the precision parameter v is beta over (.8,1.2)

BC 9.3

```
model;
{
beta[1]~dnorm(.8,1)
 beta[2]~dnorm(-.8,1)
v~dbeta(.8,1.2)
for( i in 1:20){
l[i]<-1+step(i-k-10)}
punit[1]<-.02
punit[2]<-.02
punit[3]<-.02
punit[4]<-.02
punit[5]<-.02
```

```
punit[6]<-.02
punit[7]<-.02
punit[8]<-.02
punit[9]<-.09
punit[10]<-.25
punit[11]<-.25
punit[12]<-.09
punit[13]<-.03
punit[14]<-.03
punit[15]<-.02
punit[16]<-.02
punit[17]<-.02
punit[18]<-.02
punit[19]<-.02

for( i in 1:20){mu[i]<-0}
Y[1,1:20]~dmnorm(mu[],tau[,])

for( i in 1:20){Sigma[i,i]<-v*(1+pow(beta[l[i]],2))}
for ( i in 1:19){Sigma[i,i+1]<-v*beta[l[i]]}
for ( i in 1:18){for ( j in i+2: 20) {Sigma[i,j]<-0}}
for ( i in 2:20){Sigma[i,i-1]<-v*beta[l[i]]}
 for( i in 3:20){ for ( j in 1:i-2 ){Sigma[i,j]<-0}}
 tau[1:20,1:20]<-inverse(Sigma[,])
k~dcat(punit[])
}
list(Y=structure(.Data=c( 1.26747046, 1.59926817, −0.30936599, −2.71169235,
−0.64682899, 0.85501113, −0.05213715, 0.93088400, 1.57629016, 1.25087828,
−0.62645381, 0.68480637, −0.98254327, 2.26378369, −0.94671687, −1.08407460,
1.14380376, 0.3483814, 0.93311517, −1.717707426 ),.Dim=c(1,20)))
list(v=1, beta=c(.8,−.8), k=10)
```

Based on the simulation generated with BC 9.3, the results over the posterior analysis are reported in Table 9.4.

TABLE 9.4

Posterior Analysis MA(1) Series One Shift

Parameter	Value	Mean	SD	Error	2½	Median	97½
β_1	0.8	1.241	.5883	.008581	.2679	1.86	2.494
β_2	−0.8	−1.228	.4366	.004718	−2.215	−1.17	−.4684
k	10	1.751	.226	.0338	1	1	11
v	1	.6278	.199	.002008	.2468	.6318	.9662

The posterior distribution for the shift point k is surprising in that it does not identify the actual shift point value $m=10$. This is not surprising because of the small sample size for each regime. The student will be asked to devise a model where an increase in the number of observations in each regime would indeed improve the posterior mean as an estimate of the actual shift point.

9.5 Changing Models in Econometrics

In this section, Bayesian statistical inferences on structural shift in economics will be explored. Various theories in economic development and growth assume the economic relationship changes over time. With the introduction of regression analysis as the principal tool of data processing, in the 1950s and 1960s, attempts were made to describe changes of economic associations. For example, Sengupta and Tintner[13] use long-run per capita U.S. data over the period 1869–1953 and fitted a logistic curve to each of the four equal subperiods to show that the upper asymptote of the income curve followed an increasing sequence from one subperiod to the other. The upward shift of the upper asymptote from one subperiod to the other is explained in a descriptive way, as opposed to testing whether such a shift is significant. See also Broemeling and Tsurumi[12, pp. 8–10] for additional information. Structural change in a regression framework is defined as a switch in the regression equation, that is to say, one or more regression coefficients change. Chow[14] proposed an F-test for the situation for two regression phases and the join point that separates observations into two subsamples is known. By the 1970s, the F-test proposed by Chow had been extensively used in empirical studies to test for structural change. As long as it is known to which the observations belong (the shift point is known), the F-test can be applied to either two time series or cross-sectional data. A historical time point or some threshold value of a key variable may serve as the known shift point. For example, Hamermesh[15] estimated the wage equations that switch phase at the threshold value of the price index.

Ever since Goldfield[16] examined the stability of the U.S., demand-for-money equation, the stability of the money demand equation has been accomplished in two ways: (1) examination of post-sample forecast performances by descriptive or inferential statistical criteria and (2) tests of hypotheses of parameter shifts. He examined the stability of the money demand equation by post-sample period forecasts of 10 quarters beginning with the first quarter of 1976 and concluded that it showed no tendency in forecasts errors up to 1973; however, beginning in 1974, the forecasts began to overpredict real money balances in a substantial way. These conclusions are based on descriptive measures such as the root-mean-square errors of forecast and the signed magnitude of the forecast error in each period.

The statistical tests of prediction performance of the money demand equation are based on a series of normalized errors in one-period-ahead predictions from the sequentially extended sample periods. For more information about the history of structural change in econometrics, see Broemeling and Tsurumi.[12]

9.6 Regression Model with Autocorrelated Errors

There are many ways to model structural change in a regression model with autocorrelated errors. For example, the regression parameters could change from one regime to the next, or the autocorrelation parameter could change or the variance of the errors could change. Perhaps, all three could change, and as can be seen, there are many scenarios that allow for change in a regression model.

First to be considered is a shift in the regression parameters with one unknown shift point:

$$Y(t) = X(t)\beta_1 + e(t), t = 1, 2, ..., m$$
$$Y(t) = X(t)\beta_2 + e(t), t = m + 1, 2, ..., n \tag{9.22}$$

where β_1 and β_2 are distinct $p\times1$ regression parameters, $X(t)$ is a $1\times p$ vector of known independent variables, $Y(T)$ is the observation at time t, and the errors follow a first-order autoregressive process

$$e(t) = \theta e(t-1) + \varepsilon(t), t = 1, 2, ..., n \tag{9.23}$$

where the $\varepsilon(t) \sim nid(0, \tau), t = 1, 2, ..., n$ and $\tau > 0$ is an unknown precision parameter. Also, the autocorrelation parameter is assumed unknown and $m = 1, 2, ..., n-1$. The model is rewritten as:

$$Y(t) = \theta Y(t-1) + [X(t) - \theta X(t-1)]'\beta_1 + e(t), t = m + 1, 2, , n, Y(m+1)$$
$$= \theta Y(m) + [X(m+1)\beta_2 - \theta X(m)\beta_1] + e(m+1)$$
$$= Y(t-1) + [X(t) - \theta X(t-1)]'\beta_2] + e(t), t = m + 2, ..., n \tag{9.24}$$

where $X(0)$ and $Y(0)$ are known initial values of the independent and dependent variables, respectively.

Consider the assignment of the prior distribution to the unknown parameters. Suppose that (β_1, β_2, τ) and θ are independent of the shift point m, which has a uniform distribution over the range $1, 2, ..., n-2$. In addition, suppose the conditional prior density of $\beta = (\beta_1, \beta_2)$ given τ is a 2p-dimensional normal with mean $\mu = \begin{pmatrix} \mu_1 \\ \mu_2 \end{pmatrix}$ and $2p{\times}2p$ precision matrix $Q\tau$, where Q is positive definite.

Now suppose the marginal prior distribution of τ is gamma with parameters α and β, and finally suppose the prior distribution of θ is uniform. Based on the above information, it can be shown that:

1. The marginal posterior mass function of m is

$$f(m|Y) = \int_R |H(\theta)|^{-1/2} C(\theta)^{-(2\beta+n)/2} d\theta \qquad (9.25)$$

where

$$C(\theta) = 2\beta + \mu'Q\mu + \sum_{t=1}^{t=n} [Y(t) - \theta Y(t-1)]^2 - \tilde{\beta}'H(\theta)\tilde{\beta}, \qquad (9.26)$$

$$H(\theta) = \left[\frac{Q_{11} + Z1Z_1 + \theta X'(m)X(m), \; Q_{12} - \theta X(m)X'(m)}{Q_{21} - \theta X'(m+1)X(m), \; Q_{22} + Z2Z_2 + X(m+1)X'(m+1)} \right], \qquad (9.27)$$

and

$$Q = \begin{pmatrix} Q_{11}, Q_{12} \\ Q_{21}, Q_{22} \end{pmatrix}. \qquad (9.28)$$

In addition,

$$\begin{aligned} Z_1 &= [X(1) - \theta X(0), ..., X(m) - \theta X(m-1)], \\ Z_2 &= [X(m+2) - \theta X(m+1), ..., X(n) - \theta X(n-1)], \end{aligned} \qquad (9.29)$$

$$\tilde{\beta}_1 = H_{11.2}^{-1}(\theta)[\alpha_1 - H_{12}(\theta)H_{22}^{-1}(\theta)\alpha_2], \qquad (9.30)$$

$$\tilde{\beta}_2 = H_{22.1}^{-1}(\theta)[\alpha_2 - H_{21}(\theta)H_{11}^{-1}(\theta)\alpha_1],$$

$$H_{11.2}(\theta) = H_{11}(\theta) - H_{12}(\theta)H_{22}^{-1}(\theta)H_{21}(\theta), \qquad (9.31)$$

$$H_{22.1}(\theta) = H_{22}(\theta) - H_{21}(\theta)H_{11}^{-1}(\theta)H_{12}(\theta),$$

and

$$H(\theta) = \begin{bmatrix} H_{11}(\theta), H_{12}(\theta) \\ H_{21}(\theta), H_{22\cdot}(\theta) \end{bmatrix}. \qquad (9.32)$$

Finally,

$$\alpha_1 = Q_{12}\mu_2 + Q_{11}\mu_1 + Z_1V_1 - \theta[Y(m+1) - \theta Y(m)]X(m)$$

$$V_1 = [Y(1) - \theta Y(0), ..., Y(m) - \theta Y(m-1)]. \qquad (9.33)$$

and
$$V_2 = [Y(m+2) - \theta Y(m+1), ..., Y(n) - \theta Y(n-1)].$$

Therefore, computing the marginal posterior mass function of m is quite complicated and involves the numerical integral (9.25).

2. The marginal posterior density of β is

$$g(\beta|Y) = \sum_{m=1}^{m=n-1} \int_R |H(\theta)|^{-1/2} C(\theta)^{-(2\alpha+n)/2} f(\beta, \theta) d\theta, \beta \in R^{2p}, \tag{9.34}$$

where $f(\beta, \theta)$ is a t density on β with $2\alpha + n$ degrees of freedom and precision matrix $(2\alpha + n)H(\theta)C^{-1}(\theta)$. Thus, the marginal posterior distribution of β is a mixture of t densities where the mixing distribution is that of θ.

3. The marginal posterior density of τ is

$$g(\tau|Y) = \sum_{m=1}^{m=n-1} \int_R |H(\theta)|^{-1/2} \tau^{(2\alpha+n)/2-1} e^{-C(\theta)/2} d\theta, \tau\rangle 0. \tag{9.35}$$

When a shift point is used to model structural change in this chapter, one is assuming a priori that a change will occur and will occur exactly one time. For additional information about structural change in regression models, See Salazar, Broemeling, and Chi.[17]

The following example is due to Salazar[18] and allows for smooth changes in a regression model with autocorrelated errors. Consider ψ as a nonnegative function that satisfies the following three conditions:

1. $\psi(0) = 0$,
2. $\psi(n_t) = 1$, *where* $n_t = s_t/\gamma$
   $\lim n_t \to \infty$ (9.36)
3. $s_t = 0, t \leq t^*$
   $s_t = t - t^*, t > t^*$

where t^* is a fixed number.

ψ is called a transition function and γ the transition parameter, which measures the nature of the change beginning at time t^*.

Consider a regression model with autocorrelated errors, with p independent variables $X_1(t), ..., X_p(t)$ and p regression coefficients

$\beta_1, \beta_2, \ldots, \beta_p$ and that β_1 changes from $\beta_{10} + \beta_{11}$ beginning at time t^*, then a time t

$$Y(t) = X_1(t)\beta_{10} + X_1(t)\psi(s_t/\gamma)\beta_{11} + X_2(t)\beta_2 + \\ \cdots + X_p(t)\beta_p + e(t). \tag{9.37}$$

where

$$e(t) = \theta e(t-1) + \varepsilon(t) \tag{9.38}$$

and $e(t) \sim nid\,(0, \tau > 0)$.

Note that $X_i(t)$ is the value of the i-th independent variable at time t, $Y(t)$ the value of the dependent variable at time t, θ the autocorrelation of the first-order error process and τ the precision of the noise of the AR(1) error process. Therefore, (9.37) represents a model where one of the regression coefficients changes beginning at time t^*. It is quite difficult to assign a proper prior distribution to the parameters, but for now consider the following: let the prior distribution for the parameters of (9.37)

$$\xi(\beta, t^*, \gamma, \theta, \tau) \propto \tau^{a-1} e^{-b\tau}, \tau > 0, \beta \in R^p, \gamma > 0, \theta \in R \tag{9.39}$$

Then, one may demonstrate that the corresponding posterior density of (t^*, γ) is

$$\xi(t^*, \gamma | data) \propto \int_R |Z'Z|^{-1/2}(b + v'Qv)^{-(n-p-1)/2} d\theta \tag{9.40}$$

where

$$Z = (Z_1 - \theta Z_0, \ldots, Z_n - \theta Z_{n-1}),$$

$$Z_t = [X_1(t), X_1(t)\psi(s_t), X_2(t), \ldots, X_p(t)]',$$

$$V = [Y(1) - \theta Y(0), \ldots, Y(n) - \theta Y(n-1)],$$

and $Q = I - Z(Z'Z)^{-1}Z'$.

One sees that numerical integration is necessary in order to determine the joint posterior density of t^* and γ.

Salazar gave the following example with a gradual change in the parameters of a regression model with autocorrelated errors.

$$Y(t) = 3X(t) + \varDelta(t)X(t) + e(t), \tag{9.41}$$

$$e(t) = \theta e(t-1) + \varepsilon(t),$$

$$\varepsilon(t) \sim nid(0,\tau), \tau > 0,$$

where $\tau = 1, \theta = .5, \varDelta(t) = 0, t = 1, 2, \ldots, 18,$
$\Delta(20) = .2, \Delta(21) = .3, \Delta(22) = .5, \Delta(23) = .8, \Delta(24) = 1,$
$\Delta(25) = 1.2, \Delta(26) = 1.3, \Delta(27) = 1.4, \Delta(28) = 1.5, \Delta(29) = 1.6,$ and
$\Delta(30) = 1.7.$

The following R code generates 30 observations from the model (9.40)

RC 9.3
```
x<-c
(1,2,3,4,5,6,7,8,9,10,11,12,13,14,15,16,17,18,19,20,21,22,23,24,25,26,27,28,29,30)
for ( t in 1:19) y[t]<-3*x[t]+a[t]
y[19]<-3.1*x[19]+a[19]
y[20]<-3.3*x[20]+a[20]
y[21]<-3.3*x[21]+a[21]
y[22]<-3.5*x[22]+a[22]
y[23]<-3.8*x[23]+a[23]
y[24]<-4*x[24]+a[24]
y[25]<-4.2*x[25]+a[25]
y[26]<-4.3*x[26]+a[26]
y[27]<-4.4*x[27]+a[27]
y[28]<-4.5*x[28]+a[28]
y[29]<-4.6*x[29]+a[29]
y[30]<-4.7*x[30]+a[30]
```

The 30 observations are components of the vector Y:

Y= (0.7276652, −3.7745247, 1.3457925, −2.4380528, 6.5949287, −1.3368424, −0.9515683, −7.0050599, 7.7666499, 0.2136659, 3.6857020, 3.3086897, 7.7006381, 0.4766789, 1.6403963, 6.0240775, 6.1942768, −6.7045389, 57.6427460, 65.2289357, 68.7726750, 76.9789305, 89.0451224, 94.4202446, 104.0528631, 111.4633308, 119.9051729, 126.1763260, 132.5670994, 140.4973345)

Based on Figure 9.1, there is strong evidence that the shift point is at $t^* = 19$.
Inferences about t^* are given when the parameters are $\theta = .5, \tau = 1,$ and $t^* = 19$. Note the marginal posterior density of t^* is determined by numerical integration of (9.40) with respect to y and assuming the noninformative prior density (9.39) with $a=b=0$.
The density of the shift point t^* has the values $\xi(t)$ as shown in Table 9.5.

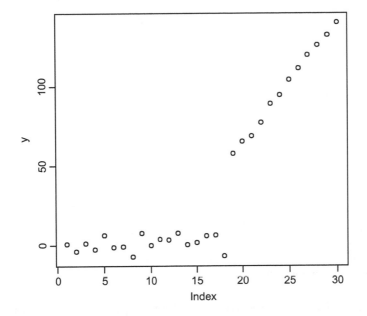

FIGURE 9.1
Regression with Autocorrelated Errors with Shift Point at $t^* = 19$

TABLE 9.5

Posterior Density of t^*

Time	$\xi\,(t)$
13	0
14	0
15	0
16	.03
17	.18
18	.28
19	.27
20	.13
21	0
22	0
23	0

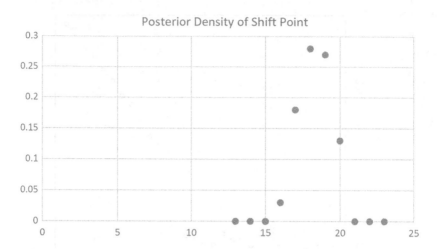

FIGURE 9.2
Posterior Density of Shift Point

Based on Table 9.5, the corresponding graph of the posterior density (9.40) of $\xi(t)$ versus time of the shift point is depicted in Figure 9.2.

The graph clearly shows the shift point t^* is in the neighborhood of $t^* = 19$. See Broemeling and Tsurumi[12, pp. 164-171] for additional details about the Bayesian analysis for a regression model with autocorrelated errors.

9.7 Another Example of Structural Change

Consider the two-phase linear regression model with AR(1) errors

$$y(t) = \beta_{11} + \beta_{12}t + w(t), t = 1, 2, \ldots, m$$
$$y(t) = \beta_{21} + \beta_{22}t + w(t), t = m + 1, \ldots, n \qquad (9.42)$$

where

$$w(t) = \theta w(t - 1) + e(t), t = 1, 2, \ldots, n, \qquad (9.43)$$

$$e(t) \sim nid(0, \tau), t = 1, 2, \ldots, n,$$

and $m = 1, 2, \ldots, n{-}1$.

Thus, there are two linear regressions with the same AR(1) process for the errors. The first regression has intercept β_{11} and slope β_{12}, while the second has intercept β_{21} and slope β_{22}.

The following R code generates 12 observations from each phase with the following parameter values: $\beta_{11} = 4, \beta_{12} = 5, \beta_{21} = 4, \beta_{22} = -5$, and $\tau = 1$.

RC 9.4

```
set.seed(1)
y1<-w<-e<-rnorm(10, sd=1)
 for ( t in 2:10) y1[t]<-4+5*t+w[t]
w[t]<-.6*w[t-1]+e[t]
y2<-v<-rnorm(10,sd=1)
 for ( t in 2:10) y2[t]<-4 -5*t+v[t]
v[t]<-.6*v[t-1]+e[t]
```

y1=(−1.101253, 14.075074, 18.601691, 26.144079, 27.806729, 32.541689, 39.213165, 42.994968, 49.673158, 54.783032)

Figure 9.3 depicts the linear regression for the first phase with slope 5
y2=(1.511781, −5.610157, −11.621241, −18.214700, −19.875069, −26.044934, −31.016190, −35.056164, −40.178779, −45.406099)

Figure 9.4 reveals the linear regression for the second phase with slope −5.

From the above two plots, the change in slope of the linear regression is obvious.

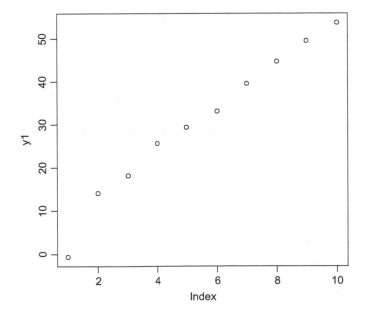

FIGURE 9.3
Linear Regression with AR(1) Errors, First Phase

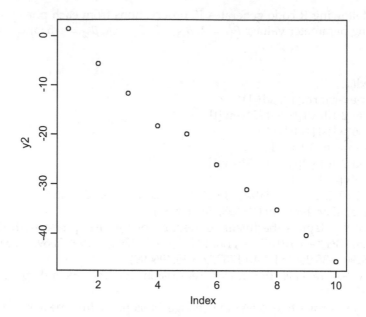

FIGURE 9.4
Linear Regression with AR(1) Errors, Second Phase

The above observations will be used for the Bayesian analysis that will estimate the two regression parameters, the autocorrelation coefficient, and the error precision.

The posterior analysis is executed with BC 9.4, assuming the prior distribution for the regression coefficients is noninformative with mean zero and precision .001. The prior mass function for the shift point is uniform over 1,2, ...9 with probability .02, at 10 with probability .1, and the remainder for .11, …, .19 is uniform with probability .0775, and is executed with 25,000 observations for the simulation and 2,000 initial values.

BC 9.4.

```
model;
{

theta~dunif(.3,.9)
beta[1]~dnorm(0,.001)
 beta[2]~dnorm(-0,.001)
v~dunif(.5,1.5)
for( i in 1:20){
l[i]<-1+step(i-k-10)}

Y[1,1:20] ~ dmnorm(mu[], tau[, ])
```

```
for( i in 1:20){mu[i]<-4+beta[l[i]]*i}

tau[1:20,1:20]<-inverse(Sigma[,])

    for(i in 1:20){Sigma[i,i]<-v/(1-theta*theta)}
    for(i in 1:20){for(j in i+1:20){Sigma[i,j]<- v*pow(theta,j)*
    1/(1-theta*theta)}}
    for( i in 2:20){ for(j in 1: i-1){Sigma[i,j]<-v*pow(theta,i-1)*
    1/(1-theta*theta)}}
 for( i in 1:9){
    s[i]<-.02}
 s[10]<-.1
 s[11]<-.1
for ( i in 12:19){ s[i]<-.0775}

k~dcat(s[])

      }
```

list(Y= structure(.Data=c(−0.6264538, 14.1836433, 18.1643714, 25.5952808, 29.3295078, 33.1795316, 39.4874291, 44.7383247, 49.5757814,53.6946116, 1.511781, −5.610157, −11.621241, −18.214700, −19.875069, −26.044934, −31.016190, −35.056164, −40.178779, −45.406099),.Dim=c(1,20)))
list(theta=.6,v=1, beta=c(5,.5),k=11)

Table 9.6 gives the details for estimating the various parameters.

The posterior analysis reveals 'good' estimates for all the parameters except the shift point *k*, which has a posterior mean of 1 very far from the true value of 10. Note the prior for *k* was almost uniform over the integers from 1 to 19. The reader will be asked to see the effect of the prior of *k* on the other parameters.

9.8 Testing Hypotheses

Consider the AR(1) process

TABLE 9.6

Posterior Analysis for Regression Model with AR(1) Errors

Parameter	Value	Mean	SD	Error	2½	Median	97½
β_1	5	4.686	.1885	.00135	4.314	4.688	5.054
β_2	−5	−1.959	.076	.000623	−2.11	−1.958	−1.809
k	10	1	0	.000000	1	1	1
θ	.6	.8971	.00287	.000031	.8893	.8893	.8999
v	1	1.481	.01886	.000204	1.486	1.486	1.5

$$Y(t) = \theta_1 Y(t-1) + e(t), t = 1, 2, ..., m$$
$$Y(t) = \theta_2 Y(t-1) + e(t), t = m+1, ..., n \qquad (9.44)$$
$$e(t) \sim nid(0, \tau), t = 1, 2, ..., n$$

where the unknown parameters are m, θ_1, θ_2, and τ.
Our goal is to test the hypothesis
$H : \theta_1 = \theta_2$ versus $A : \theta_1 \neq \theta_2$
with the Bayesian approach presented by Lee.[19] To do this, one needs the conditional density of the observations, given θ_2, θ_2, τ, and the shift point m, namely:

$$f(Y|\theta_2, \theta_2, \tau, m) = (\tau/2\pi)^{n/2} \exp(-\tau/2)$$
$$\{\sum_{t=1}^{t=m} [Y(t) - \theta_1 Y(t-1)]^2 + \sum_{t=m+1}^{n} [Y(t) - \theta_2 Y(t-1)]^2\}, \qquad (9.45)$$

where $\theta_1 \in R, \theta_2 \in R, \tau > 0$, and $m = 1, 2, ..., n-1$.
It can be shown that

$$\sum_{t=1}^{t=m} [Y(t) - \theta_1 Y(t-1)]^2 = (\theta_1 - A_1^{-1}C_1)A_1(\theta_1 - A_1^{-1}C_1) + D_1$$

$$A_1 = \sum_{t=1}^{t=m} Y^2(t-1)$$

$$C_1 = \sum_{t=1}^{t=m} Y(t)Y(t-1) \qquad (9.46)$$

$$D_1 = \sum_{t=1}^{t=m} Y^2(t)$$

and in the same way that

$$\sum_{t=m+1}^{t=n} [Y(t) - \theta_2 Y(t-1)]^2 = (\theta_2 - A_2^{-1}C_2)A_2(\theta_2 - A_2^{-1}C_2) + D_2$$

$$A_2 = \sum_{t=m+1}^{t=n} Y^2(t-1)$$

$$C_2 = \sum_{t=m+1}^{t=n} Y(t)Y(t-1) \qquad (9.47)$$

$$D_2 = \sum_{t=m+1}^{t=n} Y^2(t)$$

According to Lee, the posterior probability of the null hypothesis is

$$p_0 = \lambda_0 f(y|\theta_1 = \theta_2)/[\lambda_0 f(y|\theta_1 = \theta_2) + \lambda_1 f_1(y)] \tag{9.48}$$

where

$$f_1(y) = \int p_1(\theta_1, \theta_2, \tau) f(y|\theta_1, \theta_2, \tau) d\theta_1 d\theta_2 d\tau, \tag{9.49}$$

$$f(y|\theta_1, \theta_2, \tau, m) = (\tau/2\pi)^{n/2} \exp -(\tau/2)[(\theta_1 - A_1^{-1}C_1) \\ A_1(\theta_1 - A_1^{-1}C_1) + (\theta_2 - A_2^{-1}C_2)A_2(\theta_2 - A_2^{-1}) + D_1 + D_2], \tag{9.50}$$

$$f(y|\theta_1 = \theta_2 = \theta, m) = [\Gamma(n/2)2^{n/2}/(2\pi)^{n/2}(D_1 + D_2)^{n/2}]/E(m), \tag{9.51}$$

and

$$E(m) = \{1 + [(\theta - A_1^{-1}C_1)A_1(\theta - A_1^{-1}C_1) + (\theta - A_2^{-1}C_2) \\ A_2(\theta - A_2^{-1})]/D_1 + D_2\}^{n/2}, \tag{9.52}$$

where $\lambda_0 + \lambda_1 = 1$, λ_0 is the prior probability of the null hypothesis, and $p_1(\theta_1, \theta_2, \tau)$ is the prior density of the parameters under the alternative hypothesis. Note that care must be taken when using this formal Bayesian approach to testing hypotheses.

The data above with 10 observations in each phase of the change point model will be used to test the hypothesis. Also, it will be assumed that m is known with the value $m=10$. These data values are: For the first regime, the data are elements of the vector y1
y1=(−1.101253, 14.075074, 18.601691, 26.144079, 27.806729, 32.541689, 39.213165, 42.994968, 49.673158, 54.783032).

For the second regime, the information is contained in the vector y2:
y2=(1.511781, −5.610157, −11.621241, −18.214700, −19.875069, −26.044934, −31.016190, −35.056164, −40.178779, −45.406099)

There is now sufficient information to compute the posterior probability (9.48) of the null hypothesis, and computation of the probability p_0 of the null hypothesis is left as exercise for the student.

9.9 Analyzing Threshold Autoregression with the Bayesian Approach

A Bayesian approach to the analysis of threshold autoregressive models is presented based on a conditional likelihood function, and the posterior density of the threshold is found numerically. As for the remaining parameters, they are expressed as mixtures where the mixing distribution is that if the threshold, and the posterior means and standard deviations are obtained by formula. However, the Bayesian credible intervals are found with simulations of the posterior distribution which are easily achieved by random number generation from known distributions. A much studied area of time series is that of nonlinear models and the Jones[20] book sparked considerable interest which in turn developed into research about threshold models, first studies by Tong and Lim.[21] This section of the book is largely taken from Cook and Broemeling[22] and the reader is referred to this source for additional references. This area of time series has a vast literature, and the Bayesian approach to threshold autoregression appears to begin with Pole and Smith,[23] which was followed by an investigation of Broemeling and Cook[24] who used the Bayesian approach to estimate the parameters in a threshold auto-regression, where each regime was an AR(p) series, where p is an integer.

The following describes a two-regime threshold autoregressive series, where the order of the process is p and the delay parameter is d:

$$Y(t) = \theta_{10} + \sum_{j=1}^{j=p} \theta_{1j}Y(t-j) + e(t), Y(t-d) \leq r \tag{9.53}$$

and

$$Y(t) = \theta_{20} + \sum_{j=1}^{j=p} \theta_{2j}Y(t-j) + e(t), Y(t-d) > r,$$

where $t = 0, \pm 1, \pm 2, \ldots, r \in R,$

$$\theta_1 = \begin{bmatrix} \theta_{10} \\ \theta_{11} \\ . \\ . \\ \theta_{1p} \end{bmatrix}, \tag{9.54}$$

and

$$\theta_2 = \begin{bmatrix} \theta_{20} \\ \theta_{21} \\ . \\ . \\ \theta_{2p} \end{bmatrix} \tag{9.55}$$

Also, the error process $e(t) \sim nid\,(0, \tau > 0), t = 0, \pm 1, \pm 2, \ldots$. It is assumed the delay parameter d is known, and that if $Y(t-d) \leq r$, the series follows the first AR(p) process with parameter θ_1. A very simple threshold model will now be described.

Let

$$i_1(t,r) = 1, Y(t-d) \leq r \qquad (9.56)$$

and

$$i_1(t,r) = 0, Y(t-d) > r$$

and

$$i_2(t,r) = 1 - i_1(t,r). \qquad (9.57)$$

In addition, let define the following matrices be either the identity matrix or a matrix of all zeros. Also, let
$I_j(t,r) = i_j(t,r)I_{p+1}$, where I_{p+1} is the identity matrix of order $p+1$. Now, let $I(t,r) = [I_1(t,r), I_2(t,r)]$ be the $p+1$ by $2(p+1)$ matrix.

And finally let $Z'(t) = [1, Y(t-1), Y(t-2), \ldots, Y(t-p)]$, where if the series is to have a zero mean, the first component of $Z'(t)$ can be deleted. This is sufficient information to define the TAR(2,p,d) two-regime threshold model as

$$Y(t) = Z'(t)I(t,r)\theta + e(t) \qquad (9.58)$$

where $I(t,r)$ selects θ_1 or θ_2 in the appropriate manner. We are now in a position to explain the prior and posterior distributions.

It is easy to write matrix expressions for the sequence of observations for the TAR(2,p,d) time series as

$$\begin{bmatrix} Y(p+1) \\ Y(p+2) \\ \cdot \\ \cdot \\ Y(n) \end{bmatrix} = \begin{bmatrix} Z'(p+1)I(p+1,r) \\ Z'(p+2)I(p+2,r) \\ \cdot \\ \cdot \\ Z'(n)I(n,r) \end{bmatrix} \theta + \begin{bmatrix} e(p+1) \\ e(p+2) \\ \cdot \\ \cdot \\ e(n) \end{bmatrix}, \qquad (9.59)$$

which can be simplified to

$$Y = H(r)^\theta + e,$$

and the conditional likelihood function (conditional on the first p observations) written as

$$f(Y|Z(p), \theta, \tau, r) \propto \tau^{(n-p)2} \exp(-\tau/2)[Y - H(r)\theta]'[Y - H(r)\theta], \ \tau > 0, \theta \in R^{2p+2},$$
$$(9.60)$$

What prior distribution should be assigned to the parameters? Suppose we have little knowledge, then one could assign as a prior the density

$$\xi(\theta, \tau, r) = \xi_1(\theta|\tau)\xi_2(\tau)\xi_3(r) \propto 1/\tau, \tau > 0, \theta \in R^{2p+2}, r \in D_r, \qquad (9.61)$$

where $D_r = \{r, r = Y_{(i)}, i = 1, 2, \ldots, n - 1\}$ and $Y_{(i)}$ is the i-th-order statistic.

The posterior density of the parameters is the product of the likelihood function (9.60) and the prior density (9.61), namely

$$\zeta(\theta, \tau, r|data) \propto \tau^{(n-p)/2-1} \exp(-\tau/2)[Y - H(r)\theta]'[Y - H(r)\theta], \qquad (9.62)$$

where $\tau > 0, \theta \in R^{2p+2}$, and

$$Y = \begin{bmatrix} Y(p+1) \\ Y(p+2) \\ \cdot \\ \cdot \\ Y(n) \end{bmatrix}.$$

The posterior density (9.62) will be simplified by letting

$$\theta(r) = [H'(r)H(r)]^{-1}H'(r)Y, \qquad (9.63)$$

$$wss(r) = Y'Y - \theta'(r)H'(r)H(r)\theta(r), \qquad (9.64)$$

and

$$P(r) = [v/wss(r)]H'(r)H(r), \qquad (9.65)$$

where $v = n - 3p - 2$.
Now the posterior is written as

$$\xi(\theta, \tau, r|Y) \propto \tau^{(n-p)/2-1} \exp\{(\tau/2)wss(r) + [\theta - \theta(r)]'H'(r)H(r)[\theta - \theta(r)]\} \qquad (9.66)$$

and, conditional on r is the density of a normal-gamma distribution where

$$(\theta|\tau, r, z(p)) \sim normal\left\{\theta(r), [\tau H'(r)H(r)]^{-1}\right\}, \tag{9.67}$$

where $z'(t) = [1, Y(t-1), Y(t-2), \dots, Y(t-p)]$.
Also,

$$(\tau|r, Y, z(p)) \sim gamma(v/2, wss(r)/2),$$

and the conditional density of r is

$$\xi(r|Y, z(p)) \propto |H'(r)H(r)|^{-1/2}[wss(r)]^{-v2}. \tag{9.68}$$

Finally, since the conditional distribution of (θ, τ) is normal-gamma, the marginal posterior distribution of θ is $T_{2(p+1)}[v, \theta(r), P(r)]$, a multivariate t of dimension $2(p+1)$, with v degrees of freedom, mean $\theta(r)$, and precision $P(r)$.

9.10 A Numerical Example of Threshold Autoregression

This section presents a numerical example comparing quadrature and Monte–Carlo approaches to analyzing the various posterior distributions. The *Mathematica* programming language is involved in making the comparison and the example is the two-phase process

$$\begin{aligned}
Y(t) &= \theta_{10} + \theta_{11}Y(t-1) + e(t), Y(t-d) \le r \\
Y(t) &= \theta_{20} + \theta_{22}Y(t-1) + e(t), Y(t-d) < r
\end{aligned} \tag{9.69}$$

where $\theta_{10} = 1$, $\theta_{11} = 0.5$, $\theta_{20} = -1$, $\theta_{22} = -0.3$, $r=0$, and $d=1$. Also, the $e(t) \sim N(0,1)$

The startup values used to initiate the data generation were the mean of the first process which is 2, and this initial value induces a transient. In order to eliminate the transient, 200 data values are generated and the first 100 discarded, and thus, the resulting dataset consists of 100 values. The Monte–Carlo approach used 5,000 random variables from the distribution of $(\theta, \tau, r|Y, Z(1))$ as follows:

(1) A value of r is generated from $\xi(r|y, Z(1))$
(2) A value τ is generated from $\xi(\tau|y, Z(1))$
(3) A value θ is generated from $\xi(\theta|r, \tau Y, Z(1))$.

This algorithm was repeated 5,000 times with both $n=20$ and $n=100$ observations, and the posterior analysis for r is reported in Table 9.7.

As the sample size increases from 20 to 100, the posterior mean and median become closer to $r=0$.

The following R command generates 100 values from the model (9.69)

y=tar.sim(n=100,Phi1=c(1,0.5),   Phi2=c(0.1,−0.3),p=1,d=1,sigma1=1,thd= −1, sigma2=2[1] −2.12673392

The 100 values are components of the vector y.

y=c(0.99136731−0.60359972, 1.95290338, 0.73039125, 4.90424052, 0.26380232, −0.13198270, 0.97717077, −2.31589458, −0.14342346, −1.39803061, 1.88643857, −2.63837841, −2.94280322, −1.24140249, 1.64821880, −3.04703142, −0.11012470, 0.02308353, 0.51476591, −1.71630146, −0.41211066, −1.95545682, 0.67518661, −0.40778079, 0.35315843, 0.44870330, −2.82951335, −0.67631871, −0.51618352, 1.82178215, −5.63495535, −0.93255076, 1.26197052, 0.85541279, −2.29206070, 1.13255940, −0.46858948, 1.59597106, 3.22079702, −2.66782231, 0.92523632, −3.95885172, −0.93900261, 3.74907881, 0.25912544, −0.54315224, 2.80871670, −2.91631956, −0.80859553, 2.86331323, 1.85014583, −2.55438769, −0.38055795, −0.24616848, 1.34521828, −0.56555601, 0.74952627, 2.82808388, −0.40484278, −0.90786044, −2.47712040, −0.58812382, 2.46090764, −0.02995577, −2.55316402, −0.55726869, −1.95138871, −0.11807526, 0.62878300, −2.13636435, 0.49949003, −6.37864312, −0.85348787, −3.65569062, −0.46970901, 2.26934466, −2.20710007, −0.40050552, −0.47058389, 3.11636998, −0.84418824, −2.65769098, −1.39125205, 0.51881297, −0.70612012, 0.21062101, −1.90537074, 0.93022427, −1.44439213, 0.81818753, −1.64701280, 0.40840722, 0.14174609, −0.10742983, −1.18622028, 0.21433897, 1.33196227, 1.24950114)

The corresponding figure portrays the above values generated from the TAR model (9.69). See Figure 9.5.

Another example of a TAR model is based on the book of Cryer and Chan[25, pp. 396–398]

Consider the TAR model

TABLE 9.7

Posterior Distribution of r

	$N=20$	$N=100$
Mean	.125	−.073
Variance	.057	.003
SD	.238	.056
.10 quantile	−.075	−.108
.90 quantile	4.76	−.075
Median	.186	−.075

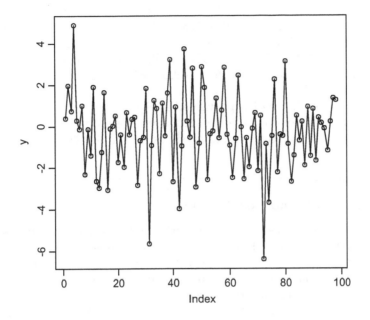

FIGURE 9.5
100 Values Generated from TAR Model (9.69)

$$
\begin{aligned}
Y(t) &= 0.5Y(t-1) + e(t), Y(t-1) \le -1 \\
Y(t) &= -1.8Y(t-1) + e(t), Y(t-1) > -1
\end{aligned}
\tag{9.70}
$$

The following R command generates 100 observations from the above two-phase TAR model using the TSA package.

y=tar.sim(n=100,Phi1=c(0,0.5),   Phi2=c(0,–1.8),p=1,d=1,sigma1=1,thd=–1, sigma2=2)

Y<-c(–6.61563254,   –3.22586884,   –2.92941542,   –0.65995618,   3.42853203,
–4.39732470, –1.55602278, 0.72285555, –1.66320287, 0.06082296, 0.39074377,
–0.06728621, 0.46990598, –2.19558659, –1.12149297, 0.33006120, –3.48371297,
–1.49555154, –1.17108279, 0.22283128, –1.51897891, 0.57730421, –5.18654510,
–3.58224847, –0.79431759, –1.03101841, –0.94350304, –0.28138086, –1.14331123,
–1.83191345, –0.94184987, –0.68447546, 2.53305468, –8.20520614, –4.44663277,
–1.62530269, 1.59447830, 0.91889598, 0.26085158, –1.85891177, –0.90206876,
3.47694920, –6.14596271, –2.85833887, –0.66807346, 3.30291163, –5.11276593,
–1.31164688, –1.03128753, –1.66236790, 0.11668228, –0.18687841, –0.75477746,
1.98086794, –2.25186310, –2.84636889, –0.94727434, 5.18251989, –7.58908299,
–5.42018222, –1.50697474, 0.01840941, 1.74687777, –3.2499147,–1.47745262,
–0.2017039, –2.48629964, –1.55774318, –0.29964253, 2.11544701, –4.59490462,

−3.87445109, −1.47112758, −0.29380608, 4.30402349, −7.05088044, −4.22397698, −1.02865127, −0.64783582, −1.99037215, −0.26476824, 1.83391621, −3.15051824, −1.51956308, −1.03223754, −0.89600915, 3.57948262, −7.89122989, −4.71492450, −2.39614815, 0.29236682, 0.16851671, 2.13237692, −1.66141137, 1.27780757, −1.81310221, −1.36371864, 1.23906527, −9.20758121, −4.85034568)
plot(Y, type = "o").

The corresponding plot is portrayed in Figure 9.6.

Based on the above data generated from the model (9.70), the student will be asked to do a Bayesian analysis using WinBUGS.

9.11 Comments and Conclusions

This chapter is focused on times series with parameters that shift. In general, there are three types of shift point problems: (a) those models whose parameters abruptly change value after a given time point, (b) those that gradually change over a given time interval, and (c) those that change when the immediate past observations fall below some threshold value. The subject matter is illustrated with many examples, where R generates the appropriate number of observations, and the analysis is executed with a Bayesian approach often employing WinBUGS.

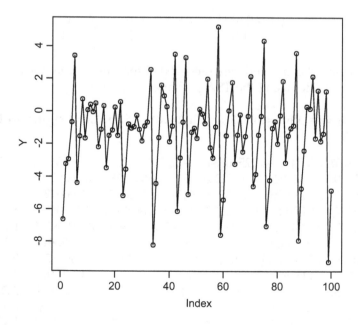

FIGURE 9.6
Simulation of TAR Model 9.6

9.12 Exercises

1. (a) Using RC 9.1, generate 25 observations from the first phase of model (9.20).

   (b) Generate 25 observations from the second phase of model (9.20).

   (c) Use BC 9.1 with 45,000 observations for the simulation with 5,000 initial values and execute a posterior analysis using data that generated with RC 9.1.

   (d) Check your results with that reported in Table 9.2.

2. Derive the posterior mass function (9.17) of the shift point.

3. (a) Based on (9.21) as a model for the shift point, use RC 9.2 to generate 25 observations for each phase of the model.

   (b) Based on the 50 observations generated with RC 9.2, use BC 9.3 to execute the Bayesian analysis. Employ 45,000 observations for the simulation with 5,000 starting values.

   (c) Compare your results with those reported in Table 9.4. They should be quite similar!

4. (a) Describe the model (9.22) and show it is equivalent to (9.24).

   (b) Based on the prior distribution assigned to the parameters, derive the marginal posterior mass function (9.25) of the shift point.

   (c) Derive the marginal posterior density of β. What is the mixing distribution?

5. (a) Use RC 9.3 to generate 30 observations from the model (9.40) and plot these observations using the R code. Your plot should be similar to Figure 9.1.

   (b) Does the plot reveal the value of the shift point t^*. What is it?

   (c) Verify Table 9.5, the posterior density of t^*.

6. (a) Describe the model (9.42), a two-phase regression model with AR(1) errors.

   (b) Assuming $\beta_{11} = 4, \beta_{12} = 5, \beta_{21} = 4, \beta_{22} = -5, \tau = 1$, use RC 9.4 to generate 40 observations, with 20 for each phase.

   (c) Plot the 20 observations for each of the two phases.

   (d) Using these 40 observations as data and BC 9.4, execute a posterior analysis with 45,000 observations for the simulation and 4,000 starting values.

   (e) How do your results compare to those reported in Table 9.6?

7. (a) Describe the two-phase AR(1) model (9.44), and consider a test of the hypothesis $H: \beta_1 = \beta_2$ versus the alternative $A : \beta_1 \neq \beta_2$

(b) Show the conditional density of the observations given the parameters are given by (9.45).

(c) Given the information from formulas (9.45)–(9.48) and the 10 observations generated for each phase, and assuming the shift point is $m=10$, compute the posterior probability p_0 given by (9.48) of the null hypothesis. Also, assume the prior probability of the null hypothesis is ½.

(d) Based on the posterior probability of the null hypothesis, would you reject it?

8. (a) Provide a complete description of the TAR model defined by (9.59).

(b) Derive the likelihood function given by (9.60).

(c) Based on the prior (9.61) and the likelihood function (9.60), derive the posterior density (9.62) of the TAR parameters.

9. (a) Describe the two-phase TAR model given by (9.67) with the parameters $\theta_{10} = 1, \theta_{11} = .5, \theta_{12} = -1, \theta_{22} = -0.3, r = 0, d = 1$.

(b) Describe the Monte–Carlo algorithm that generates 5,000 values for the simulation of the parameters of the TAR model.

(c) Verify the posterior analysis of r reported in Table 9.7 for $n=20$ and $n=100$.

(d) What is the posterior median of r for $n=20$.

(e) What is the posterior mean for r for $n=100$.

(f) Does $n=100$ observations provide a 'better' estimate of r compared to the estimate of r based on $n=20$.

10. (a) Refer to (9.68) and the associated TAR model. Using the indicated R command, generate 100 observations from the TAR Model (9.68).

(b) Plot the above observations using the R command plot(y, type = 'o') and compare your results to Figure 9.6.

(c) Based on the 100 observations generated in part (a) above, execute a Bayesian analysis with WinBUGS using 35,000 observations for the simulation and 5,000 starting values.

References

1. Page, E.S. (1954). "Continuous Inspection Schemes", *Biometrika*, 41, 100–114.
2. Page, E.S. (1957). "On problems in which change in parameter occurs at an unknown time point", *Biometrika*, 44, 248–252.

3. Chernoff, H. and Zacks, S. (1964). "Estimating the current mean of a normal distribution which is subjected to changes over time", *Annals of Mathematical Statistics*, 35, 999–1018.

4. Kander, Z. and Zacks, S. (1966). "Test procedures for possible changes in parameters of statistical distributions occurring at unknonwn time points", *Annals of Mathematical Statistics*, 37, 1196–1210.

5. Chin Choy, J. and Broemeling, L.D. (1980). "Some Bayesian inferences for a changing linear model", *Technometrics*, 22, 71–78.

6. Booth, N.B. and Smith, A.F.M. (1982). "A Bayesian approach to retrospective identification of change points", *Journal of Econometrics*, 19, 7–22.

7. Broemeling, L.D. (1974). "Bayesian inferences for a changing sequence of random variables", *Communications in Statistics*, 3(3), 243–255.

8. Broemeling, L.D. (1977). "Forecasting future values of a changing sequence", *Communications in Statistics*, A6(1), 87–102.

9. Broemeling, L.D. (1985). *Bayesian Analysis of Linear Models*. Marcel-Dekker Inc., New York.

10. Shao, X. and Zhang, X. (2010). "Testing for change points in time series", *Journal of the American Statistical Association*, 105(491), 1228–1240.

11. Barberi, M.I. and Conglinani, C. (1998). "Bayesian analysis of autoregressive time series with change point", *Journal of the Italian Statistical Association*, 7, 243.

12. Broemeling, L.D. and Tsurumi, H. (1987). *Econometrics and Structural Change*. Marcel Dekker Inc., New York.

13. Sengupta, J.K. and Tintner, G. (1963). "On some aspects of trend in the aggregate models of economic growth", *Kylos*, 16, 47–61.

14. Chow, G. (1960). "Tests for the equality between two sets of coefficients in two linear regressions", *Econometrica*, 28, 561–605.

15. Hamermesh, D.S. (1970). "Wage bargains, threshold effects, and the Philips", *Quarterly Journal of Economics*, 84, 501–517.

16. Goldfield, S.M. (1976). "The case of the missing money", *Brookings Papers on Economic Activity*, 3, 588–635.

17. Salazar, D., Broemeling, L.D. and Chi, A. (1981). "Parameter changes in a regression model with autocorrelated errors", *Communications in Statistics*, A (17), 1751–1758.

18. Salazar, D. (1982). "Structural changes in time series models", *Journal of Econometrics*, 19, 147–164.

19. Lee, P.M. (1997). *Bayesian Statistics, An Introduction, Second Edition*. John Wiley & Sons Inc., New York.

20. Jones, D.A. (1978). "Non-linear autoregressive processes", *Proceedings of the Royal Society*, A360, 711–795.

21. Tong, H. and Lim, K.S. (1980). "Threshold autoregression, limit cycles, and cyclical data", *Journal of the Royal Statistical Society*, B42, 245–292.

22. Cook, P. and Broemeling, L.D.(1996). "Analyzing threshold autoregressions with a Bayesian approach", *Advances in Econometrics*, 11B, 89–107.

23. Pole, A.M. and Smith, A.F.M. (1985). "Bayesian analysis of some threshold switching models", *Journal of Econometrics*, 29, 17–119.

24. Broemeling, L.D. and Cook, P. (1992). "Bayesian analysis of threshold autoregressions", *Communications in Statistics – Theory and Methods*, 21(9), 2459–2482.

25. Cryer, J.D. and Chan, K.S. (2008). *Time Series Analysis: with Applications in R*. Springer, New York.

10

Residuals and Diagnostic Tests

10.1 Introduction

This is the last chapter of the book which will emphasize the diagnostic tests that check the validity of the model proposed at hand, and the tests are based on the residuals of the fitted model, which should appear as independent with mean zero and constant precision, if indeed the tentatively proposed model is indeed the 'true' one.

For example, suppose the model is an AR(1)

$$Y(t) = \theta Y(t-1) + e(t), t = 1, 2, \ldots \tag{10.1}$$

where the errors are

$$e(t) = Y(t) - \theta Y(t-1), t = 1, 2, \ldots \tag{10.2}$$

and $e(t) \sim nid(0, \tau), t = 1, 2, \ldots$. Of course, in the Bayesian approach, each of the $e(t)$ has a posterior distribution induced by the posterior distribution of θ and τ. In addition to the AR(1) model, various models in the previous chapters will be the basis for the illustration of Bayesian diagnostic tests. In order to detect deviations of the residuals from their optimal form, Bayesian analytical techniques and plots of the residuals over time are also examined as part of the diagnostic test mechanism. Residual analysis for time series will include the autoregressive series, regression models with autoregressive errors, moving average models, and regression models with moving average errors.

10.2 Diagnostic Checks for Autoregressive Models

Consider the AR(1) series(10.1) with $\theta = 0.6$ and $\tau = 1$, where 50 values are generated with RC 10.1, namely

RC 10.1

```
set.seed(1)
y<-w<-rnorm(50)
for( t in 2:100) y[t]<-.6*y[t-1]+w[t]
time <- 1:100
plot(time,y)
acf(y)
pacf(y)
```

The first 50 values are labeled by the vector y= (−0.62645381, −0.19222896, −0.95096599, 1.02470121, 0.94432850, −0.25387129, 0.33510628, 0.93938847, 1.13941444, 0.37826027, 1.73873733, 1.43308564, 0.23861080, −2.07153341, −0.11798913, −0.11572708, −0.08562651, 0.89246030, 1.35669738, 1.40791975, 1.76372922, 1.84037383, 1.17878928, −1.28207813, −0.14942113, −0.14578142, −0.24326436, −1.61671100, −1.44817665, −0.45096443, 1.08810089, 0.55007281, 0.71771530, 0.37682414, −1.15096507, −1.10557361, −1.05763412, −0.69389387, 0.68368905, 1.17338918, 0.53950991, 0.07034427, 0.73916994, 1.00016516, −0.08865660, −0.76068912, −0.09183151, 0.71343402, 0.31571420, 1.07053625)

Using the 50 values as data, the posterior analysis is executed with BC 10.1 using 5,000 initial values and 35,000 for the MCMC simulation. Note the prior distribution for θ is $n(0,.001)$ and for τ is gamma(.01,.01), both noninformative. The posterior distribution of the standardized residuals is induced by the posterior distribution of the precision and autoregressive parameter.

BC 10.1

```
model;
{
mu[1]<-theta1*ym1
y[1]~dnorm(mu[1],tau)
for ( t in 2:50){ mu[t]<-theta1*y[t,1;;1,-1]
y[t]~dnorm(mu[t],tau)
}
# prior distributions
theta1~dnorm(0.0,.001)
tau~dgamma(.01,.01)
ym1~dnorm(0.0,.001)
# the residuals
for( i in 2:50){ R[i]<-y[i]-mu[i]}
# standardized residuals
for( I in 2:50){(r[i]<-R[i]−.1373)/.8273
}
list(y=c(−0.62645381, −0.19222896, −0.95096599, 1.02470121, 0.94432850, −0.25387129, 0.33510628, 0.93938847, 1.13941444, 0.37826027, 1.73873733,
```

1.43308564, 0.23861080, −2.07153341, −0.11798913, −0.11572708, −0.08562651, 0.89246030, 1.35669738, 1.40791975, 1.76372922, 1.84037383, 1.17878928, −1.28207813, −0.14942113, −0.14578142, −0.24326436, −1.61671100, −1.44817665, −0.45096443, 1.08810089, 0.55007281, 0.71771530, 0.37682414, −1.15096507, −1.10557361, −1.05763412, −0.69389387, 0.68368905, 1.17338918, 0.53950991, 0.07034427, 0.73916994, 1.00016516, −0.08865660, −0.76068912, −0.09183151, 0.71343402, 0.31571420, 1.07053625))

The results of the posterior analysis for the standardized residuals (computed as a posterior mean) are reported in Table 10.1.

Also, the mean of the residuals is $\bar{R}$ = .1373265 and standard deviation S_R =.8273, and thus, the standardized residuals are $r[i] = \{R[i] - \bar{R}\}/S_R$ which appear in Table 10.1.

The plot below depicts the values of the standardized residuals over time 1 to time 50 (Figure 10.1).

The mean of the standardized residuals is $\bar{r} = .014207$ and the standard deviation is $S_r = 1.0000938$; thus, one can conclude the AR(1) model (10.1) fits the model well; however, one should conduct a formal test of the hypothesis that the mean of the standardized residuals is indeed 1. Also, the reader should compute the quartile–quartile plot of the standardized residuals to assess how the sample quartiles deviate from those of the normal distribution.

10.3 Residuals for Model of Color Data

This example is taken from Cryer and Chan,[1, pp. 174_176] where the data is generated with the following R code.

RC 10.2
```
> win.graph(width=4.875, height=3,pointsize=8)
> data(color)
> m1.color=arima(color,order=c(1,0,0))
> m1.color
Call:
arima(x = color, order = c(1, 0, 0))
sigma^2 estimated as 24.83: log likelihood = −106.07, aic = 216.15
> plot(rstandard(m1.color),ylab='Standardized residuals',type='b')
> abline(h=0)
```
The data appear as elements of the vector Y=(67, 63, 76, 66, 69, 71, 72, 71, 72, 72, 83, 87, 76, 79, 74, 81, 76, 77, 68, 68, 74, 68, 69, 75, 80, 81, 86, 86, 79, 78, 77, 77, 80, 76, 67), and a plot of the data appears in Figure 10.2.

TABLE 10.1

Posterior Analysis of Standardized Residuals

Residual Number	Standardized Residual
1	
2	−.0265
3	−1.201
4	1.637
5	.3673
6	−1.033
7	.3898
8	.7706
9	.6538
10	−.365
11	1.711
12	.5344
13	−.7281
14	−2.812
15	.9208
16	−.2358
17	−.2008
18	.9636
19	.9443
20	.7307
21	1.13
22	1.021
23	.1667
24	−2.415
25	.4143
26	−.2535
27	−.3735
28	−1.976
29	−.9569
30	.1484
31	1.417
32	−.1468
33	.3751
34	−.1364
35	−1.781
36	−.8192
37	−.7882
38	−.377
39	1.072

(Continued)

TABLE 10.1 (Cont.)

Residual Number	Standardized Residual
40	.8466
41	−.2102
42	−.4011
43	.6858
44	.6043
45	−.8667
46	−1.033
47	.1745
48	.7509
49	−.2078
50	.9407

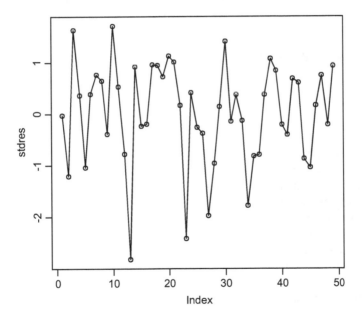

FIGURE 10.1
Standardized Residuals of AR(1)

This example is a time series from an industrial color process, where the values are the color property from consecutive batches of the process. See Cryer and Chan[1, pp. 1–3] for additional details.

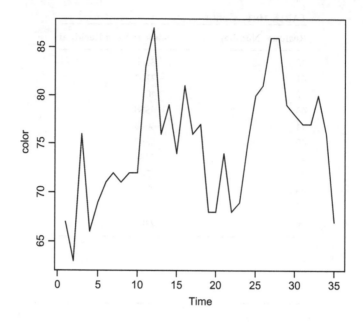

FIGURE 10.2
Values for the Model Color Data

BC 10.2
model;
{
mu[1]<-theta*ym1
y[1]~dnorm(mu[1],tau)
for (t in 2:35){ mu[t]<-theta*y[t-1]
y[t]~dnorm(mu[t],tau)
Z[t]~dnorm(mu[t],tau)
}
# prior distributions
theta~dnorm(0.0,.001)
tau~dgamma(.01,.01)
ym1~dnorm(0.0,.001)
for(i in 2:35){
R[i]<-y[i]-mu[i]}
For(I in 2:35){r[i]<-(R[i]−.161)/5.67537}
}
 list(y=c(67, 63, 76, 66, 69, 71, 72, 71, 72, 72, 83, 87, 76, 79, 74, 81, 76, 77, 68, 68, 74, 68, 69, 75, 80, 81, 86, 86, 79, 78, 77, 77, 80, 76, 67))
list(theta=.5, tau=1,ym1=22)
    The posterior analysis is reported in Table 10.2.

TABLE 10.2

Posterior Analysis for Color Data

Parameter	Mean	SD	Error	2½	Median	97½
R[2]	−3.862	.8887	.004576	−5.615	−3.867	−2.11
R[3]	13.13	.8355	.004303	11.48	13.13	14.78
R[4]	−9.843	1.008	.005191	−11.83	−9.838	−7.856
R[5]	3.135	.8753	.004508	1.409	3.141	4.862
R[6]	2.143	.9151	.004713	.3365	2.147	3.946
R[7]	1.147	.9416	.004849	−.7117	1.153	3.003
R[8]	−.8512	.9549	.004918	−2.736	−.8463	1.031
R[9]	1.147	.9416	.004849	−.7117	1.152	3.003
R[10]	.1488	.9549	.004918	−1.736	.1537	2.031
R[11]	11.5	.9549	.004918	.9264	11.15	13.03
R[12]	4.171	1.101	.005669	1.999	4.177	6.341
R[13]	−10.82	1.154	.005942	−13.1	−10.81	−8.546
R[14]	3.157	1.008	.005191	1.168	3.162	5.144
R[15]	−4.387	1.048	.005396	−6.905	−4.831	−2.772
R[16]	7.153	.9814	.005054	5.216	7.158	9.087
R[17]	−4.833	1.074	.005532	−6.953	−4.827	−2.715
R[18]	1.157	1.008	.005191	−.8323	1.162	3.144
R[19]	−8.841	1.021	.005259	−10.86	−8.836	−6.828
R[20]	.1405	.9019	.004645	−1.639	.1452	1.918
R[21]	6.14	.9019	.004645	4.361	6.145	7.918
R[22]	−5.847	.9814	.005054	−7.784	−5.842	−3.913
R[23]	1.14	.9019	.004645	−.6394	1.145	2.918
R[24]	6.143	.9151	.004713	4.336	6.147	7.946
R[25]	5.155	.9947	.005123	3.192	5.16	7.116
R[26]	1.165	1.061	.005464	−.9287	1.171	3.257
R[27]	5.167	1.074	.005532	3.047	5.173	7.285
R[28]	.1777	1.141	.005874	−2.073	.1839	2.426
R[29]	−6.822	1.141	.005874	−9.073	−6.816	−4.574
R[30]	−.8368	1.048	.005396	−2.905	.8313	1.228
R[31]	−.8388	1.034	.005328	−2.881	−.8335	1.2
R[32]	.1591	1.021	.005259	−1.856	.1644	2.172
R[33]	3.159	1.021	.005259	1.144	3.164	5.172
R[34]	−3.835	1.061	.005464	−5.929	−3.829	−1.743
R[35]	?	?	?	?	?	?
τ	.03099	.007612	.0004106	.01794	.03038	.04768
θ	.9979	.01326	.0000683	.9718	.9979	1.024

It is remarkable that the posterior distribution of all the residuals is symmetric about the mean.

Using R, the mean of the residuals is $\bar{R} = .16106$ and the standard deviation of the residuals is $S_R = 5.67537$.

The standardized residuals are r=($-.7083,2.286,-1.762,.5248,.3497,.1742,$ $-.1778,.1742,-.00016,1.937,.7073,-1.934,.5285,-.88,1.233,-.8793,.1761,-1.586,$ $-.003099,1.054,-1.058,.1731,1.054,.8805,.1775,.8827,.00357,-1.23,-.1752,$ $-.1756,.000236,.5288,-.7035,-1.586)$

10.4 Residuals and Diagnostic Checks for Regression Models with AR(1) Errors

A model is linear if

$$Y(t) = \beta_0 + \beta_1 X_1(t) + \beta_2 X_2(t) + \ldots + \beta_m X_m(t) + Z(t) \tag{5.1}$$

where $Y(t)$ is the observation of the dependent variable at time t, $X_i(t)$ is the observation of the i-th independent variable at time t, and finally $Z(t)$ is the error term at time t. The errors $Z(t), t = 1, 2, \ldots, n$ are assumed to have mean zero, have a constant variance, and are autocorrelated. Our goal is to compute Bayesian inferences for the m regression coefficients β_i and the unknown parameters of the error process.

As a first example, consider the linear regression model

$$Y(t) = 50 + 3t + Z(t), t = 1, 2, \ldots, 100 \tag{10.3}$$

where the $Z(t)$ follows an AR(1) process with autocorrelation θ.

Consider RC 10.3 which will generate 100 observations from the linear regression model. The model is assumed to have a standard deviation of 10 for the Gaussian white noise of the AR(1) process and the autocorrelation coefficient is assigned the value $\theta = .6$, while the regression coefficients are assumed to be 3 for the slope and fifty for the intercept. The first fifty values generated by RC 5.1 appear below the code. The fifty values of the dependent variable are the components of the vector Y.

RC 10.3

```
> set.seed(1)
> u<-w<-rnorm(100,sd=10)
> for ( t in 2:100) u[t]<-.6*u[t-1]+w[t]
> time<-1:100
> y<-50+3*time+u
> plot(time, y)
```

Y=(46.73546, 54.07771, 49.49034, 72.24701, 74.44328, 65.46129, 74.35106, 83.39388, 88.39414, 83.78260, 100.38737, 100.33086, 91.38611, 71.28467, 93.82011, 96.84273, 100.14373, 112.92460, 120.56697, 124.07920, 130.63729, 134.40374, 130.78789, 109.17922, 123.50579, 126.54219, 128.56736, 117.83289, 122.51823, 135.49036, 153.88101, 151.50073, 156.17715, 155.76824, 143.49035, 146.94426, 150.42366, 157.06106, 173.83689, 181.73389, 178.39510, 176.70344, 186.39170, 192.00165, 184.11343, 180.39311, 190.08168, 201.13434, 200.15714, 210.70536)

A plot of the simple linear regression dependent variables versus time over fifty days appear in Figure 10.3, and the linear trend is obvious. What is the posterior mean of the intercept, slope, and autocorrelation?

It appears the observation begins at 50 starting at time 0.

A goal of the Bayesian analysis is to estimate the regression coefficient and the autocorrelation θ. In the statements of **BC 10.3**, beta0 is the intercept and beta1 the slope of the regression model with autocorrelation θ. The 20 observations is a vector with a multivariate normal distribution with mean vector consisting of 20 values of 50+3t, t=1,2, ..., 20, and a 20-by-20 precision matrix which is the inverse of the variance–covariance matrix of an AR(1) error process.

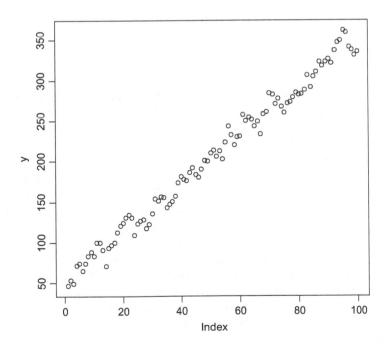

FIGURE 10.3
Simple Linear Regression

BC 10.3
model;
{

```
v~dgamma(.1,.01)
theta~dbeta(6,4)
beta0~dnorm(0,.001)
beta1~dnorm(0,.001)

    Y[1,1:20] ~ dmnorm(mu[], tau[, ])


    for( i in 1:20){mu[i]<-beta0+beta1*i}


    tau[1:20,1:20]<-inverse(Sigma[,])
    for(i in 1:20){Sigma[i,i]<-v/(1-theta*theta)}
    for(i in 1:20){for(j in i+1:20){Sigma[i,j]<- v*pow(theta,j)*
    1/(1-theta*theta)}}
    for( i in 2:20){ for(j in 1: i-1){Sigma[i,j]<-v*pow(theta,i-1)*
    1/(1-theta*theta)}}
    # the residuals
    for ( i in 2:20){ R[i]<-Y[1,i]-mu[i]}
# the standardized residuals
for ( i in 2:20){r[i]<-(R[i]-.1373)/.8273
    }
```

list(Y= structure(.Data=c(46.73546, 54.07771, 49.49034, 72.24701, 74.44328, 65.46129, 74.35106, 83.39388, 88.39414, 83.78260, 100.38737, 100.33086, 91.38611, 71.28467, 93.82011, 96.84273, 100.14373, 112.92460, 120.56697, 124.07920),.Dim=c(1,20)))
list(theta=.6,v=100, beta0=50, beta1=3)

The posterior analysis for the AR(1) regression is executed by BC 10 using 45,000 observations for the simulation and 5,000 for the initial run, and the results that display the posterior median of the residuals appear in Table 10.3. Note that the posterior analysis has to be run twice, the first time for the residuals (which allows one to compute the posterior means of the residual and the standard deviation of the residuals), then once one knows the mean and standard deviation of the residuals, one is able to compute the posterior distribution of the standardized residuals!

The vector of standardized residuals is r<-c((−.5878,−.1697, −1.042,1.043,.9048,−.4429,.1426,.7447,.9094,.03448,1.454,1.072,−.2714,-2.822, −.7604,−.8095,−.8286,.1778,.6263,.6321), where the mean is $\bar{r} = .00344$ and standard deviation is $S_r = 1.00026\,2$ and the R command plot(r,type='0') produces the graph of the standardized residuals shown in Figure 10.4.

TABLE 10.3

The Standardized Residuals for
the Linear Regression Model

Number	Residual
1	−.5878
2	−.1697
3	−1.042
4	1.043
5	.9048
6	−.4428
7	.1426
8	.7447
9	.9094
10	.03448
11	1.454
12	1.072
13	−.2714
14	−2.822
15	−.7604
16	−.8095
17	−.8286
18	−.1778
19	.6283
20	.6321

The mean of the standardized residuals is $\bar{r} = .00344$ and the standard deviation is $S_r = 1.000262$. On the basis that the mean is close to 0 and the standard deviation is extremely close to 1, one may conclude that the regression model (10.3) with AR(1) errors provides a good fit to the data. One should test the hypothesis that the mean of the residuals is 0 and its standard deviation is 1. This will be left as an exercise for the student. One could also construct a q–q plot to see the association between quartiles of the residuals and the quartiles of the normal distribution.

10.5 Diagnostic Tests for Regression Models with Moving Average Time Series

Consider the following model:

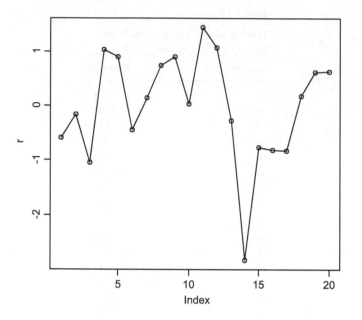

FIGURE 10.4
Standardized Residuals for Linear Regression Model

$$Y(t) = \gamma_0 + \gamma_1 t + \gamma_2 t^2 + Z(t), t = 1, 2, \ldots, n \qquad (10.4)$$

where the errors follow the MA(1) process

$$Z(t) = W(t) + \beta W(t-1), \qquad (10.5)$$

with $\gamma_0 = 1$, $\gamma_1 = 2$, $\gamma_2 = 3$, $\beta = .8$, and the variance of white noise is $\sigma^2 = 1$.

The following R code generates 20 observations from the regression process (10.4) with the known parameters values as given above.

RC 10.4

```
set.seed(1)
b<-c(.8)
y<-x<-w<-rnorm(20)
for ( t in 2:20){
for ( j in 1:1) x[t]<-w[t]+b[j]*w[t-j]}
{for( t in 1:20)y[t]<-1+2*t+3*t^2+x[t]}
```

Y=(5.373546, 16.682480, 33.311286, 57.926778, 87.605732, 120.443138, 161.831054, 210.128268, 263.166441, 321.155237, 387.267470, 458.599268,

533.690634, 614.288308, 705.353171, 801.855011, 901.947863, 1009.930884), 1123.576290, 1242.250878)

Of course, the goal of the Bayesian analysis is to estimate the unknown regression parameters $\gamma_i, i = 1, 2, 3$, the moving average coefficient β, and Gaussian noise variance σ^2. Using the above 20 values generated by RC 9.12 according to the model (6.6) and (6.7), **BC 10.4** is executed with 35,000 observations for the simulation and 5,000 for the burn-in. The 20 observations of vector Y are normally distributed with mean vector given by (6.6) and variance–covariance matrix appropriate to the MA(1) time series errors parameter β.

BC 10.4

```
model;{
beta~dnorm(.8,1)
v~dgamma(.1,.1)
g0~dnorm(1,1)
g1~dnorm(2,1)
g2~dnorm(3,1)
for( t in 1:20){mu[t]<-g0+g1*t+g2*t*t}
Y[1,1:20]~dmnorm(mu[],tau[,])
for( i in 1:20){Sigma[i,i]<-v*(1+pow(beta,2))}
for ( i in 1:19){Sigma[i,i+1]<-v*beta}
for ( i in 1:18){for ( j in i+2: 20) {Sigma[i,j]<-0}}
for ( i in 2:20){Sigma[i,i-1]<-v*beta}
  for( i in 3:20){ for ( j in 1:i-2 ){Sigma[i,j]<-0}} tau[1:20,1:20]<-inverse
(Sigma[,])
# the residuals
for(i in 2:20){ R[i]<-Y[1,i]-mu[i]}
# the standardized residuals
for(i in 2:20){r[i]<-(R[i]-.0000894)/1.10106}
}
list(Y=structure(.Data=c(5.373546,    16.682480,    33.311286,    57.926778,
87.605732,  120.443138,  161.831054,  210.128268,  263.166441,  321.155237,
387.267470,  458.599268,  533.690634,  614.288308,  705.353171,  801.855011,
901.947863, 1009.930884, 1123.576290, 1242.250),.Dim=c(1,20)))
list(v=1, g0=1,g1=2,g2=3, beta=.8)
```

The posterior analysis is reported in Table 10.4.

The standardized residuals r[i] given by the posterior mean are elements of the vector r: r<-c(−.2969,−.6583,.7832,1.373,−.6201,−.2979,.8486,.8502, −.1027,.8714,1.135,−.6371,−.6371,−2.859,−1.026,.2948,−.5738,.2727,.811,.4656). Using the R code

Plot(r, type= 'o') a plot of the standardized residuals appear in Figure 10.5.

TABLE 10.4

Posterior Analysis for Regression Model with MA(1) Errors

Parameter	Value	Mean	SD	Error	2½	Median	97½
β	.8	1.184	.5182	.01273	.375	1.182	2.318
γ_0	1	.9514	.733	.02000	−.4733	.9477	2.393
γ_1	2	2.026	.1849	.00681	1.622	2.029	2.381
γ_2	3	3.001	.0093	.00033	2.982	3.001	3.018
σ^2	1	.7069	.411	.00804	.1643	.637	1.679

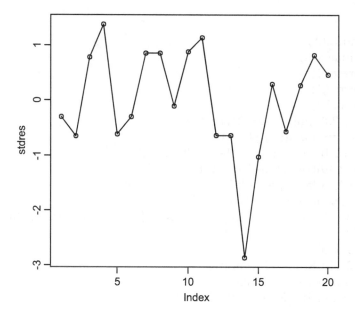

FIGURE 10.5

Standardized Residuals for Regression Model with MA(1) Errors

The mean of the standardized residuals is $\bar{r} = .0017$ and standard deviation $S_r = .98402$. What does this imply about the goodness of fit of the regression model?

To answer that question, consider a quantile–quantile plot to assess the normality of the residuals with the R command qqnorm(r) to give Figure 10.6.

This graphs shows that the standardized residuals do indeed deviate from the normal distribution (because if the residuals are normal, one would expect the plot to be linear). Another way to assess normality is to

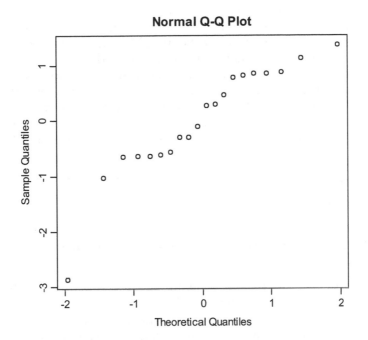

FIGURE 10.6
q–q Normal Plot of Standardized Residuals of Regression Model

graph the histogram of the standardized residuals. This is left as an exercise for the student.

10.6 Comments and Conclusions

The chapter begins with a preview of the chapter. AR(1) time series are considered where RC 10.1 generates the data from the AR(1) model followed by the Bayesian analysis that computes the residuals, and the standardized residuals are depicted in Figure 10.1.

Section 10.3 deals with the color data modeled by an AR(1) process, where RC 10.2 generates the data from the model. Based on this data, the posterior analysis is executed with BC 10.2 that produces the posterior mean of the standardized residuals depicted in Figure 10.2. Section 10.4 deals with a simple linear regression model with AR(1) errors and RC 10.3 generates the data for this model (intercept 50 and slope 3), and based on these data, the Bayesian analysis BC 10.3 computes the posterior means of the standardized residuals reported in Table 10.3 and whose values are portrayed in Figure 10.4. Section 10.5 deals with a quadratic regression model with MA(1) errors, where RC 10.4 generates data from this model and, on the basis of this data, BC 10.4 provides the posterior mean for the standardized residuals depicted in Figure 10.5. A q–q plot reveals that the standardized residuals are not normally distributed.

The topic of residuals analysis and diagnostic check for mode fit is a voluminous subject. For additional information, see Durbin and Watson,[2]-[3,4] Box and Pierce,[5] and Mcleod.[6] In the context of the dynamic linear model, see Petris, Petrone, and Campagnoli[7] and for the residual analysis of binary data from a Bayesian perspective, see Albert and Chib.[8]

10.7 Exercises

1. Write a two-page essay that summarizes the content of Chapter 10.
2. Refer to Section 10.2.
   a. Use R code 10.1 to generate 50 observations from the AR(1) series with parameters $\theta = 0.6$ and $\tau = 1$.
   b. Using the 50 values as data, execute the posterior analysis via BC 10.1 with 35,000 for the simulation and 5,000 initial values.
   c. Compute the 50 residuals R[i} via BC 10.1, and compute their mean and standard deviation.
   d. Compute the standardized residuals r[i] via BC 10.1.
   e. Compute the mean and standard deviation of the standardized residuals.
3. Refer to Section 10.4.
   a. Refer to (10.3), the linear regression model with intercept 50 and slope 3. Note that the AR(1) errors with parameters $\tau = 1$ and $\theta = 0.6$.
   b. Use RC 10.3 to generate 50 observations from (10.3).
   c. Use R to plot these 50 values and compare with Figure 10.3.
   d. Using the first 20 values generated with RC 10.3, execute the Bayesian analysis via BC 10.3 with 35,000 for the simulation and 5,000 initial values. Note the prior distribution of the parameters specified by the WinBUGS code of BC 10.3.
   e. Via BC 10.3, find the 20 residuals R[i] and compute their mean and standard deviation.
   f. Via BC 10.3, find the standardized residuals r[i] and compute their mean (which should be very close to 0) and standard deviation (which should be very close to 1).

g. Determine the normal q–q plot of the standardized residuals. Does it appear linear? If not, what does this imply how well the model fits the data?

4. Refer to Section 10.5.
   a. Use RC 10.4 to generate 20 observations from the quadratic regression series (10.4), where the errors follow a MA(1) process. Note that the parameter values used for the simulation are $\gamma_0 = 1, \gamma_1 = 2, \gamma_2 = 3, \beta = 0.8, \tau = 1$.
   b. Based on the data generated by RC 10.4, execute the posterior analysis for the above parameters with BC 10.4 using 25,000 observations for the simulation and 5,000 for the initial values.
   c. Refer to the WinBUGS code of BC 10.4, and generate the 20 residuals, R[i],i=1,2, ..., 20, and calculate their mean and standard deviation.
   d. Using the above mean and standard deviation of the residuals, refer to the WinBUGS code and generate the standardized residuals r[i], i=1,2, ..., 20.
   e. Compare your results of the posterior analysis with that reported in Table 10.4.
   f. Plot the standardized residuals r[i], i=1,2, ... 20.
   g. Compute the mean (should be close to 0) and standard deviation (should be close to 1) of the standardized residuals.
   h. Execute a normal q–q plot of the standardized residuals and compare with Figure 10.5. Does the model fit the data? Why?

References

1. Cryer, J.D. and Chan, K.S. (2008). *Time Series Analysis: with Applications in R.* Springer, New York.
2. Durbin, J. and Watson, G.S. (1950). "Testing for serial correlation in least-squares regression I", *Biometrika*, 37, 409–429.
3. Durbin, J. and Watson, G.S. (1951). "Testing for serial correlation in least-squares regression II", *Biometrika*, 38, 1–19.
4. Durbin, J. and Watson, G.S. (1971). "Testing for serial correlation in least-squares regression III", *Biometrika*, 58, 409–428.
5. Box, G.E.P. and Pierce, D.A. (1970). "Distribution of residual correlation in autoregressive-integrated moving average time series models", *JASA*, 65, 1509–1526.
6. Mcleod, A.I. (1978). "On the distribution of residual autocorrelations in Box-Jenkins models", *Journal of the Royal Statistical Society: Series A*, 40, 296–302.
7. Petris, P., Petrone, S. and Campagnoli, P. (2009). *Dynamic Linear Models in R.* Springer, London and New York.
8. Albert, J. and Chib, S. (1995). "Bayesian residual".

Index